Praise for *Restoring the Soil*

Hundreds of thousands of smallholder farmers in many countries are using, experimenting with and sharing green manure/cover crops, thereby reducing fertilizer use while improving the sustainability of crop production systems. Restoring the soil by using these technologies to feed food-insecure people will have a tremendous impact on improving global food security. Roland Bunch brilliantly shows how green manure/cover crops can best be used to achieve these objectives.

—Ing. Elmer López Rodríguez
Minister of Agriculture
Government of Guatemala

Roland Bunch's book on green manure/cover crops could not have come at a better time. Decades of experimentation and experience by farmers, technicians and scientists are presented clearly and practically, making this work an essential contribution not just for world food security but for the food sovereignty of the 2 billion smallholders who produce the bulk of the world's food.

—Dr. Eric Holt-Giménez
Executive Director
Food First/Institute for Food and Development Policy
Author of Food Rebellions: Crisis and the Hunger for Justice and other books

Green manure/cover crops are among the best solutions to sustainably increase production while improving soil fertility for smallholder farmers…. Green manure/cover crops are proving to be an effective, locally appropriate and low-external-input solution to [the coming world food] crisis.

—Dr. Jim Cornelius
Executive Director
Canadian Foodgrains Bank

Roland Bunch's book distills for us decades of rich field experience in countries all around the world, where millions of farmers have given practical meaning to the concept of synergy. Farmers are doing this by growing a great number of crops in intelligent and profitable association with a huge variety of green manure/cover crop plants. These marvelous plants enrich the soil at the same time that they enhance crop yields, conserve soil moisture, and save farmers' labor. Thanks to Bunch's keen observations, systematic analysis and critical thinking, this agroecologically-based strategy is now accessible to anyone who will read, digest and apply what is in this book. It fills a huge gap in our published knowledge about how to make the most of our land, labor and water for meeting our food needs in sustainable ways.

—Dr. Norman Uphoff
Professor of Government and International Agriculture, Cornell University
Director, Cornell International Institute for Food and Agricultural Development (CIIFAD)
Author of Reasons for Success and other books

A timely and extremely valuable book from one of the world's leading experts on green manure/cover crops. Clearly written, it is both practical and analytical, and richly illustrated with photographs from a great number of agroecosystems worldwide. Highly recommended.

—Dr. Jules Pretty
Deputy Vice-Chancellor
University of Essex
Author of Regenerating Agriculture and other books

This is the book we have been waiting for. *Restoring the Soil* distills and condenses a lifetime of learning and a wealth of experience. It is a wonderful source of knowledge and advice about a vital and neglected area of huge potential, a treasure trove for smallholder farmers around the world and those who work with them. Only Roland Bunch with his lifetime of learning and his unique worldwide experience could have written this. *Restoring the Soil* should have a major impact, widening farmers' choices and enhancing their sustainable production. It is practical, accessible and wide-ranging across an astonishing variety of green manure/cover crops. Let me hope that it will be widely available in several languages, and widely used.

—Dr. Robert Chambers
Director
Institute of Development Studies
University of Sussex
Author of Rural Development, Putting the Last First, and other books

Restoring the Soil

How to Use Green Manure/Cover Crops to Fertilize the Soil and Overcome Droughts

Second Edition

By Roland Bunch

Restoring the Soil: How to Use Green Manure/Cover Crops to Fertilize the Soil and Overcome Droughts

Edited by Dawn R. Berkelaar and Dr. Timothy N. Motis

Cover design by Jeremy Bunch

All the photos in this book were taken by Roland Bunch, unless otherwise indicated. They were taken either in smallholder farmers' fields, kitchens or markets and, except in the case of Conrado's experiment (photo 5), show practices regularly used by the owners of those fields.

First edition published by Canadian Foodgrains Bank

Second edition published by ECHO, Inc.

ISBN: 978-1-946263-30-8

For further resources, including networking with other agricultural and community development practitioners, please visit ECHO's website: www.ECHOcommunity.org. ECHO's general information website can be found at: www.echonet.org.

The cover photo, from the Valle de Angeles township in Honduras, shows a good deal of what we can achieve with green manure/cover crops. First, the farmer is proudly showing us in his right hand the two large ears of maize he harvests per plant now, instead of the one medium-sized ear per plant he used to harvest. Because he has almost tripled his productivity, he is no longer forced to spend months each year cutting sugarcane on a plantation; he now harvests more than enough maize for his family on his own land. The very healthy maize crop, as well as the black, organic-matter-laden soil at his feet, show the impact of his green manure/cover crop, mucuna (*Mucuna* spp.), which is climbing up the maize behind him. The contour grass barrier behind him and the small ditch he is standing in show he is also using soil conservation practices that are keeping that rich, black soil on his farm. Finally, his son, who is sitting on top of the 40-cm terrace face that has gradually formed behind the grass barrier, is in his late teens but is still living at home rather than having gone to find one of those almost non-existent jobs in the slums of Tegucigalpa. By making this choice, he is providing concrete evidence that he believes the farm is now going to sustainably support a fulfilling life for him and his future family.

This book is dedicated to all those anonymous smallholder farmers who,
without their realizing it or anyone else's recognizing them for it,
became the pioneers of a worldwide agricultural movement
that is handsomely benefiting all of humankind,
as well as our precious environment.
We give you our heartfelt
thanks.

Photo 1. Honduras. Increased soil fertility. The contrast between the upper half of this field and the lower half was brought about by two previous years of intercropping mucuna with the maize on the upper half.

Photo 2. Honduras. Increased soil fertility. In the same field, one can clearly see that the yield in the upper half of the maize field (behind the woman, who is the owner of the field) will easily be more than four times that of the lower half. This represents a larger improvement than we would normally expect, but is still within the range of possibilities.

Photo 3 by Tiana Ratsimbazafy. Madagascar. Increased resistance to drought. The control plot here, with no previous plantings of Lojy Be cowpeas, and the experimental plot in Photo 4, with six previous years of Lojy Be use, are about 10 m apart. The Lojy Be cowpea is a 9-month cowpea that produces both leaves and grain that are edible and widely eaten. The two photos were taken the same day. The same two clouds seen to the right of this photo had moved over to the left by the time the second photo on the facing page was shot. A very small application of the same quantity of synthetic fertilizer was applied initially to both plots. The dark green foliage of the experimental plot can be seen in this photo on the extreme right, behind the control plot.

Even though the maize here is planted much less densely than in the experimental plot, the maize in this field has tightly curled leaves, a sign of extreme drought stress. Even the Lojy Be, although drought tolerant enough to still be quite green, has only formed small clumps of vegetation rather than covering all the soil.

Photo 4 by Tiana Ratsimbazafy. Madagascar. Increased resistance to drought. The dark green color of the maize here indicates that soil fertility has improved, while its wide flat leaves show that it has suffered virtually no drought stress. The only indication that a drought occurred are the curled leaves of two hills of maize in the center of the photo where the maize had to compete for moisture with the previously established vetiver grass seen there. The Lojy Be has completely covered the ground without climbing or bothering the maize. It has thereby maintained moisture in the soil both by covering the soil and by increasing its OM content. Outside the field, the soil is still the same dry, whitish soil seen all through the control plot.

The increased fertility here—resulting in more than a tripling of the productivity of the maize, in addition to a four-month-long harvest of high-protein cowpeas—is consistent with that achieved by many other gm/ccs over six years. The increased resistance to drought is also consistent with that achieved by a good number of other gm/ccs over five or six years.

TABLE OF CONTENTS

Part II: Choosing the Best Gm/cc System for Each Situation

PREFACE

During the last 35 years, green manure/cover crops (gm/ccs) have become recognized widely as an important agricultural technology for the developing world. Millions of farmers use traditional gm/cc systems, and millions more now use systems developed and introduced during the last three decades. Books on tropical soil management written before the 1980s seldom mentioned green manures, except perhaps to remind readers that they had failed in the developing world. However, in the 1980s, gm/ccs began appearing regularly in the literature. Today, many books on sustainable agriculture in the tropics include sections on gm/ccs.

Still, when many people think of gm/ccs, they continue to picture maize fields covered by mucuna or jack beans (*Canavalia ensiformis*). Researchers frequently organize experiments with three or four gm/cc species used in inauspicious ways, and if none of them produces results, conclude that gm/ccs as a whole are worthless. In fact, over 100 gm/cc species are currently cultivated around the world, in over 150 known cropping systems. Farmers grow gm/ccs together with all the world's basic subsistence crop species, as well as with most species of vegetables, root crops and tree crops.

Faced with so many promising possibilities, even well-informed agronomists find it difficult to choose the best gm/cc system for a given situation. This happens in part because, unfortunately, some 20 different factors must be considered in selecting the gm/cc systems that will have the greatest potential in any given situation. These factors include everything from local food habits, current market conditions, the dominant cropping systems and the major weeds, to local economic needs, environmental conditions and farmer preferences. Furthermore, the best gm/cc systems almost always include not just one, but several species of gm/cc.

With this book, I hope to accomplish two things:

Part I describes general issues surrounding gm/ccs. It explains what can be expected of gm/ccs, their advantages and disadvantages, some common misconceptions about them, and the mistakes most often made in using them. It clarifies the connections between gm/ccs and soil fertility; between gm/ccs and the unique potential of agriculture in the tropics; between gm/

ccs and conservation agriculture (CA) and climate-smart agriculture (CSA); between gm/ccs, carbon sequestration and the environment; between gm/ccs and the resistance of crops to drought; and between gm/ccs, synthetic fertilizer, animal manure and composting. This information is crucial for anyone dealing with soil management.

Part II seeks to simplify the daunting task confronting programs of trying to decide which gm/cc systems might be best for a specific situation. To make this job easier, most of Part II is organized as a decision tree with commentary. Also, at the beginning of a number of the chapters, we have placed a plain decision tree to help readers see more easily the issues in the chapter and their relationship to each other. A much wider understanding of gm/ccs in general can be obtained by reading all of Part II. Nevertheless, those readers involved in specific programs can use these decision trees to proceed directly to the systems that would likely work best in their program area, without reading the entire second half of the book.

The gm/cc systems described in this book in individual text boxes have each been given a number, preceded by a capital S, for "system." The 120 numbered text boxes are divided into the following categories:

1. **Proven.** To qualify for this category, these systems have succeeded in fulfilling two criteria. The first is rapid or widespread adoption. A particularly dramatic case of widespread spontaneous adoption would be that of the 3rd maize/mucuna system in Mexico and Central America (S52). In less than 90 years, this system has spread spontaneously to the extent that an estimated 25,000 farmers presently use it. The second criterion is sustainability. Each of these systems has been used by at least 100 farmers over a period of at least five years in the total absence of any outside support or artificial subsidy. Sustainability is exemplified by the 1st maize/runner bean system (S97), which, in at least seven Latin American nations, has lasted from before Columbus arrived in the Americas to the present day.

2. **Good potential.** These systems have been adopted by less than 100 farmers or have been used for less than five years since they were last subsidized or promoted by some institution from outside of the village. Adoption rates are often very high, and the systems show a great deal of promise, but they have not yet met the criteria of proven systems. In many cases, the system has not yet qualified as "proven" because the program that introduced the system is either still working with the farmers or was until less than five years ago.

3. **Experimental.** These systems are still insufficiently tested, but hold out a great deal of promise, and therefore are well worth trying out on a small scale under appropriate conditions.

4. **Fallback.** These systems should only be promoted if other, better systems cannot be used in that situation, for whatever reason.

5. **NEVER.** These two systems illustrate problems or issues that are worth knowing about, but are systems that should never be introduced into new areas.

I have personally observed a large majority of these systems and have interviewed farmers who use them. To the best of my knowledge, all the systems described are still in use today. Of the systems listed in the 120 text boxes, those in 84 boxes are proven, those in another 9 have good potential, those in 14 boxes are still experimental, and those in 11 are fallback technologies. Two of the systems should never be promoted in new areas. A few of the individual boxes include several gm/cc systems, so the total number of gm/cc systems described in the book would be well over 150.

In no way is this list definitive. There are over 60 developing nations I have never visited, and undoubtedly many systems of which I have never heard within the countries I have visited. As time passes, more beneficial gm/cc species will be found and more systems will be developed. Even at 150+ systems, we see no reduction in the rate at which new systems are being discovered, developed, and adopted.

While those of us who have worked on this book believe it represents an important step forward in most people's understanding of gm/ccs, we fully recognize that it represents only a very early step in the huge task of developing and then describing all the gm/cc systems that already are used, or could potentially be used, around the world. We would be very grateful to any readers who would be willing to add to what we have written here. Please write me at rbunchw@proton.me. Thank you very much.

ACKNOWLEDGEMENTS

This book would not exist were it not for a handful of people who had the foresight to realize, 35 years ago, that over the long haul, world demand for energy would outstrip world supply. These people realized that a dramatic increase in energy prices would inevitably push up the price of nitrogenous fertilizers. Since over a billion people around the developing world depended on synthetic fertilizer, hundreds of millions would suffer the consequences.

Although petroleum prices have fluctuated wildly, these predictions were prophetic: real petroleum prices have more than doubled since the early 1970s. Real prices of N fertilizer in most villages have tripled in the same period. Today, because of the double-barreled threat of the death of natural fallowing (allowing the land to rest and grow back into forest) and of this rise in fertilizer prices, hundreds of millions of smallholder farmers are watching helplessly as their farms gradually turn into drought-prone wastelands.

Nevertheless, because of the foresight of this handful of people, we now have scores of effective, inexpensive, proven farming systems that raise yields and provide resistance to droughts in ways no synthetic fertilizer ever could—and at far less cost. Thousands of people, both farmers and professionals, have helped develop gm/cc systems over the last 30 years, but the true pioneers in this effort deserve special recognition:

- Ana Primavesi was the pioneering leader of the entire Brazilian zero tillage and gm/cc movement—the most successful in the world. Her classic book, *The Ecological Management of the Soil*, still has no rival worthy of the name.[1]
- Claudino Monegat probably did more than anyone else to push forward the early scientific experimentation with gm/ccs in Brazil. He later wrote his own classic, *Green Manuring in Southern Brazil*.[2]
- Valdemar de Freitas and Ademir Calegari have coordinated the agricultural organizations for the Brazilian states of Santa Catarina and Paraná, respectively, where most of the early gm/cc work in Brazil was done. As a result of the movement they led, more than two million farmers in Brazil now use gm/ccs on over 25 million hectares.

- Rolf Derpsch worked in Brazil for many years, and then spread gm/ccs to hundreds of thousands of farmers in Paraguay.
- Steve Gliessman and Roberto García pioneered work with jack beans on the Yucatán Peninsula in Mexico.
- After a false start in 1973, those of us in World Neighbors (WN)/ Honduras once again began experimenting with gm/ccs in 1983, in part because of an experiment carried out by a Honduran smallholder farmer, Conrado Zavala. By the mid-1980s, we began publicizing the benefits of gm/ccs. In the 2nd Edition of my book, *Two Ears of Corn* published in 1985, I introduced a section on gm/ccs by writing that, "one technology with tremendous untapped potential is that of using legumes as green manure crops."
- Also in 1983, Tony Rinaudo of World Vision/Australia began developing the farmer-managed natural regeneration of trees (FMNR) in Niger. He eventually inspired a movement that has spread trees across an estimated 24 million ha (based on satellite photos) of previously denuded landscape in at least ten African nations and beyond.
- Milton Flores became the director in 1989 of the International Cover Crop Clearinghouse (CIDICCO), founded by WN/Honduras. He ably ran CIDICCO for two decades, publishing and spreading information about gm/ccs to more than 80 countries around the world.

In addition to these pioneers, I would like to thank the members of the Green Manure/Cover Crop Taskforce organized by the Canadian Foodgrains Bank (CFGB), who recognized the need for this book. Dr. Tom Post and Dr. Wondimu Kenea from World Renew first captured the vision for its design, and Joe Trimble made the book possible by going through a tremendous amount of research in documents that only existed in hard copy and sending the results to Kenya, where I was writing this book.

Dawn Berkelaar, ECHO Inc. in Florida, did an incredible job of editing both the first and second editions of the book, assisted by Tim Motis, Bob Hargrave, Philip Bender, Carol Thiessen, Rachel Evans and John Longhurst. Significant amounts of technical information were provided for this edition by Neil Miller, Joseph Mureithi and Sebastian Scott.

Lastly, *Restoring the Soil* would not exist if Alden Braul from CFGB had not initiated the idea and helped push the first edition forward to completion.

Roland Bunch

Blantyre, Malawi

ABBREVIATIONS

CIDICCO	The International Cover Crop Clearinghouse (Spanish acronym), a non-governmental organization founded by WN/Honduras, that for 25 years disseminated information on green manure/cover crops around the world
EPAGRI	The Enterprise of Agricultural Research and Extension, which is the agricultural department of the government of the State of Santa Catarina in southern Brazil. It has recently been renamed CIRAM.
FAO	The United Nations' Food and Agriculture Organization, charged with supporting agricultural production in developing nations
FMNR	Farmer-managed natural regeneration of trees. This system consists of very easily establishing biodiverse forests in semi-arid and arid zones by managing the regrowth of sprouts from an "underground forest." This underground forest is made up of root systems of trees that have previously been cut off or burned multiple times and from seeds that have remained viable in the soil for years.
gm/cc	A green manure/cover crop. A gm/cc is defined as a species of plant, usually a legume, whether it be a tree, a bush, a vine, a crawling or a water-borne plant, that farmers use (among other reasons) to maintain or improve their soil fertility and control weeds.
ha	Hectare(s). One hectare is a piece of land 100 meters by 100 meters in size, equaling about 2.5 acres.
IAPAR	The Agronomic Institute of Paraná, which is the agricultural department of the government of the Brazilian State of Paraná
kg	Kilogram(s)
lb	Pound(s)
m	Meter(s)
N	Nitrogen
NGO	Non-governmental organization
OM	Organic matter
P	Phosphorus
SOM	Soil organic matter
sq	Square
t	Metric ton(s), equal to 1,000 kg. or about 2,240 pounds
WN	World Neighbors, an NGO based in Oklahoma, USA

PART I: BACKGROUND INFORMATION

Chapter 1: Introduction

The Definition of Green Manure/Cover Crops

Attempts were made during the colonial era to spread green manures (plants that fertilize the soil) around the developing world, but decades of work almost completely failed. This failure did not occur because green manure species could not fertilize the soil, but rather because the ways of using them were adapted neither to the tropics nor to the needs of smallholder farmers. For this reason, those of us who initiated the modern movement of green manuring took care first to understand smallholder farmers' needs, and then adapt the systems accordingly. Of course, some smallholder farmers had already developed well-adapted gm/cc systems, and we learned a good deal from these already-existing systems, too.

Because the basic idea of how best to use plants to fertilize the soil in a tropical environment has changed so completely during the last 40 years, it is important that we define exactly what we mean by the term "green manure/ cover crop," or gm/cc. The following definition, originally established by a group of the pioneers of gm/ccs in the late 1980s, is the accepted definition still today. A gm/cc is **any species of plant, usually a legume, whether it**

be a tree, a bush, a vine, a crawling or a water-borne plant, that farmers use (among other reasons) to maintain or improve their soil fertility or control weeds. Were I in a position to unilaterally update this definition, I would now amend it to read "…to maintain or improve their soil fertility, control weeds or make the soil resistant to drought."

The traditional use of the term "green manures" in temperate countries involves growing viny or bushy plants all by themselves (monocropped). They are then plowed into the soil (as one does with "manure") while they are in the flowering stage (i.e. still "green"), before the main food or cash crops are planted.

Smallholder farmers in the tropics, however, need to manage gm/ccs in entirely different ways.

First, because smallholder farmers possess very little land, they must maximize their use of it. They cannot afford to take land out of food production for several months of the growing season just to grow fertilizer. They need to grow food *at the same time, on the same land*, as they grow their fertilizer, or they need to grow the fertilizer on the same land at times when crops cannot be grown, such as during a dry season. Most commonly, smallholder farmers plant gm/ccs intercropped with their food crops. Occasionally, they plant the gm/ccs during the off season (dry season rotations) or just before the food crops are harvested (relayed). Farmers may also plant gm/ccs under other crops, either under fruit trees or as crawling species under almost any other crops.

Alternatively, they may plant gm/ccs called "fertilizer trees" so the trees can grow over their food crops. These trees are planted either as "dispersed shade" (a uniform, light shade over the field) or "intermittent shade" (a heavier shade that travels across the field, allowing all the crops to be shaded part of the day). These systems actually increase the crops' productivity, because crops all over the lowland tropics suffer from excessive heat. As a result, in total sunlight, they just stop growing for an hour or two in the middle of the day. They will produce as much as 30% more under a well-managed tree shade with an average of between 15 and 50% shade. And, of course, with global warming, this problem of excessive heat is going to get worse in the near future.

In all of the above ways, farmers fertilize their soil and grow food on the same, limited amount of land. Increasingly, they use several of these methods in the same field to add extra species of gm/ccs to their farming systems.

Second, smallholder farmers almost never cut down their gm/ccs at flowering stage. Instead, they leave the gm/ccs standing in the field until they have produced seeds. They allow the seeds to mature because they want to eat them, sell them, feed them to their animals or plant them the following growing season. Furthermore, since many species of tropical legume fix well over 100 kg of N/ha, farmers do not need more than half of what is fixed. Cutting down the plant after the seed has been harvested, such that the N in the seed is lost to the system, very rarely causes a problem.

Third, smallholder farmers almost always leave the gm/ccs' crop residues on top of the soil, rather than burying them. Leaving the residues on the surface eliminates the considerable expense of plowing them under or the tremendous amount of labor of digging them into the soil by hand. Furthermore, the organic matter (OM) on the surface forms a mulch, which in the tropics, especially, provides a whole series of additional benefits.

Fourth, smallholder farmers want their gm/cc species to do much more than just fertilize their soils. By far, the most desirable benefit is that they produce food. Almost all of us who work with gm/ccs have had the experience of showing farmers a new gm/cc seed that could be of tremendous value in fertilizing the soil, controlling weeds or feeding their animals, only to have them patiently ask: "And can we eat it?" Some gm/ccs, such as the cowpea and lablab bean, also produce edible leaves that can be cooked and eaten fresh or dried, green seeds that can be eaten like common peas (*Pisum sativa*), and/or edible pods that can be eaten in soups like snow peas.

Finally, most smallholder farmers are more interested in controlling weeds than in fertilizing their soil. That is why we have joined the concept of "cover crops" (plants that control weeds) with that of "green manures" (ones that fertilize the soil).

Therefore, while smallholder farmers are still using the basic "green manure" idea of using plants to fertilize the soil, they have totally transformed the way "green manures" are managed.

The above definition of gm/ccs obviously includes trees that fertilize the soil (also called "agroforestry species"), as well as leguminous gm/ccs whose seeds people eat (also called "grain legumes"). This use of categories that overlap, such that one species may be included in more than one category, may be a bit confusing, but is nothing new. After all, pigeon peas (*Cajanus cajan*) and lablab beans, as woody perennials whose seeds are widely consumed, both fall within the traditional agronomic categories of "agroforestry species" and of "grain legumes."

The Definition of Gm/cc Systems

We distinguish gm/cc systems from each other by the species of gm/cc that is used, by the subsistence or cash crop with which it is associated, by planting dates, or by important differences in how the systems are managed. In some cases, particularly when a gm/cc is used in a rotation with various crops or to recuperate wastelands, a single gm/cc "system" may be used to benefit any number of traditional crops. Text boxes are used throughout much of the book to distinguish descriptions of gm/cc systems from the general text. Each text box will normally describe a single system, but to avoid unnecessary repetition, one text box may occasionally include more than one of several very similar systems.

Each text box is identified by a number preceded by a capital "S." That is, "S24" refers to a particular text box in which a gm/cc system, or group of very similar systems, is described.

Chapter 2: The History of Green Manure/Cover Crops

The Role of Smallholder Farmers

For thousands of years, farmers all across the tropics kept their soils fertile by using gm/ccs in what we call traditional fallowing systems. They would leave agricultural land to grow into forests or grasslands for 10 to 15 years or more. They thereby allowed the native vegetation to grow long enough that the increase in soil organic matter (SOM) would restore the soil's natural fertility. These systems were ecologically sound—their sustainability can be measured in millennia.

To this day, the most convincing evidence we have that gm/ccs can be highly sustainable is that tens of millions of farmers maintained their soils' fertility for millennia using traditional fallows. Some scientists have argued that this is not an accurate analogy, because in today's gm/cc systems, nutrients are taken off the fields in harvests. But smallholder farmers using fallows also harvested nutrients from their fields. Much more important, the vast majority of nutrients that modern smallholder farmers using gm/ccs extract from their fields, as well as those of additional foods they buy, are returned to the fields directly or indirectly in the form of kitchen waste and human waste.

Smallholder farmers developed over 75% of the modern gm/cc systems included in this book. Furthermore, these farmer-developed systems are by far the ones most widely used to this day. Systems such as the 1st maize/runner bean (*Phaseolus coccineus*) system (S97) once used by many millions of farmers from the northern US to northern Chile, the 1st maize/cowpea system (S23) and 1st maize/rice bean (*Vigna umbellata*) system (S35) used by millions of farmers in Mexico, Central America and Vietnam, and the systems that associated the three *Vigna* species with rice, maize and cassava, used by millions of farmers in Southeast Asia—all these systems were developed because of the natural inclination of smallholder farmers to experiment and innovate.

This fact, perhaps more than any other, shows how appropriate these systems are for such farmers, and how interested they are in finding and adopting alternatives to both fallowing and synthetic fertilizers. That smallholder farmers around the world are using over 100 species of gm/ccs, and continue to discover new ones, further confirms the popularity of gm/ccs.

The Birth and Rebirth of a Modern Movement

Green manures entered recorded history almost as soon as it began. Certain Swiss farmers were using green manures as early as 4,000 BC, while Chinese authors in 500 BC wrote about how mung beans (*Vigna radiata*) could improve the soil.[3] In Europe and the United States, green manures were by far the most common method of maintaining soil fertility up until the end of World War II. Agricultural textbooks from the 1930s and earlier contained a tremendous amount of information about green manures—information that today has largely been forgotten.[4]

Since the 1940s, the effectiveness of OM in fertilizing the soil has been cast into doubt by the dominance of synthetic fertilizer in developed nations, and by the role of fertilizer companies in downplaying the value of OM and the systems that use it well. For instance, I once checked the 15 entry-level college soils textbooks at Cornell University's Mann Library. Every single one of them had at least one co-author who was a fertilizer company official. Understandingly, none of them admitted that green manures might ever act as a substitute for any amount of synthetic fertilizer.

During the 1950s and 1960s, with the successful introduction of synthetic fertilizers throughout the developed world, virtually nothing was done anywhere to promote green manures. Then, by sheer coincidence, between 1975 and 1983, three independent organizations on three different continents began the ongoing efforts with modern gm/ccs that led to today's worldwide movement. None of these organizations knew of the others' work until ten years later. And in two of the three cases (Brazil and Central America), the work was begun because program personnel became familiar with gm/cc systems that smallholder farmers had already developed in their area.

The governments of two Brazilian states became the first organizations to begin working seriously with gm/ccs. In 1975, IAPAR, the agricultural arm of the government of the State of Parana, initiated a new emphasis on zero tillage and gm/ccs. In 1978 EPAGRI (recently renamed CIRAM) did the same in the State of Santa Catarina.

Between the two of them, EPAGRI and IAPAR have introduced well over a hundred new gm/cc systems, involving over 70 gm/cc species. They also introduced many of these systems into Rio Grande do Sul and several other Brazilian states. Today, well over 2 million farmers in Brazil use gm/ccs and zero tillage (together called "Conservation Agriculture") on some 25 million hectares.[5] Using gm/ccs, minimum tillage and much less synthetic fertilizer/ha than farmers in the United States, many of these people harvest from 7 to 8 t/ha of maize. This effort is undoubtedly the most successful effort ever made of developing and disseminating gm/cc systems and of introducing zero tillage to developing world farmers.

EPAGRI's work has become so popular and its impact so widespread that once when I was in Santa Catarina during a state-wide election campaign, increased financial support for EPAGRI was a widely advertised plank in the platforms of *both* the major political parties.

World Neighbors/Central America suffered a false start in Guatemala in 1973. We tried to disseminate a traditional maize/choreque (*Lathyrus nigrivalvis*) system in San Martin Jilotepeque, but choreque had a very limited ecological niche and failed in most of the area, so we gave it up.

Photo 5. Honduras. Conrado Zavala's experiment. The maize on the left and the single row of stunted, ankle-high maize trailing off to the right were planted the same day, with the same seed. The only difference between these contrasting rows of maize consisted of an extremely generous application of compost on the left.

By 1983, however, we had learned of Steve Gliessman and Roberto García's work in Mexico with a traditional maize/jack bean system (S57) and of the traditional 3rd maize/mucuna system (S52) in northern Honduras. Furthermore, in 1982, Conrado Zavala, a smallholder farmer participant in a WN program in Guinope, Honduras, became doubtful about some of our program's more optimistic statements about OM. To find out the truth, he built 16 compost heaps and buried the compost in one part of his worn-out field (photo 5). The maize in the unfertilized one-row "control plot" barely grew 30 cm high and produced no grain. The rest of the field, heavily fertilized with compost, grew over 2 m high and yielded over 4 t/ha.

Conrado's experiment taught us two very important lessons. The good news was that the ability of OM to achieve impressive increases in productivity far surpassed anything we, or most agronomists, had ever imagined. The bad news was that to achieve such increases in yields, one had to apply 25 t/ha or more of OM (green weight). But if Conrado could produce such dramatically positive results, why were so many smallholder farmers around the world still producing so little? It was inconceivable that no smallholder farmers had ever thought of this possibility.

The answer is that, despite the startling increase in yield, Conrado's experiment was a money-loser. The cost of making, managing, transporting and spreading so much compost across his field greatly exceeded the value of his increase in maize yield, even if we included the compost's residual effect on future yields. Given the tons of OM necessary to improve soil fertility and the exorbitant cost of making, transporting and spreading so much material, we knew we had to abandon the idea of smallholder farmers' transporting most of the OM onto their basic grain fields. Compost, animal manure, kitchen garbage and sweepings from the family compound could help, especially with high-value crops. But if yields of the developing world's basic grains were to increase substantially, the vast majority of the OM would have to be produced *in situ*.

Conrado's experiment also gave evidence, later confirmed in thousands of farmer experiments around the world, that soil fertility improvement is not basically a problem of soil chemistry. Virtually any soils in the world (except those that lack manganese, have a pH under 4.5, are polluted or are too saline), can be dramatically improved, and then maintained over time, if we just add enough OM to them. Soil fertility improvement is a problem of *economics*; the real problem is how to apply such a large amount of OM to the soil at a reasonable cost.

Subsequent experience around the world has confirmed that a) soil improvement with OM does require at least 20 to 25 t/ha/year of above-ground biomass, in addition to normal crop residues; b) at those rates it can improve virtually any unpolluted, non-saline soil with a decent pH; and c) reducing costs is the major factor in achieving success, which means, among other things, that farmers must produce the bulk of the OM in their fields.

These insights convinced us that we should renew our work with gm/ccs. In 1983, WN program director Laureano Jacobo established experiments with mucuna, jack beans and lablab beans near Comayagua, Honduras. The success of these and many later experiments led WN to establish the International Cover Crops Clearinghouse (CIDICCO) in 1989. Since then, CIDICCO has provided information on gm/ccs to interested organizations in over 80 nations.

In early 1993, during a workshop I was facilitating in Chile, one of the more animated participants kept making very interesting comments. My years of working with or visiting several hundred agricultural development programs had taught me several of those comments were unique to people who were involved in exceptionally good programs. I pulled Valdemar de Freitas aside and asked him to tell me about his work. Within minutes, my hunch was confirmed, and I was planning a trip to southern Brazil to see what turned out to be one of the very best agricultural development programs I have ever visited: the EPAGRI program.

Valdemar was the director of EPAGRI, the agricultural development program of the State of Santa Catarina. During the visit, we drove for hundreds of kilometers with farms on both sides of the highways covered by gm/ccs and erosion-controlling contour grass barriers. Within a year, I had arranged for Valdemar to speak at conferences in Mexico, Honduras, India and the Philippines. EPAGRI and IAPAR staff, most notably Ademir Calegari, then began training people in Cuba, Angola, Zambia, Zimbabwe, Mozambique and elsewhere. Over time, they had a profound influence on the entire Conservation Agriculture movement that has spread across Africa.

Meanwhile, also in 1983, Tony Rinaudo began experimenting with a system of farmer-managed natural regeneration of trees (FMNR) in semi-arid and arid regions of Niger. Gradually, with the help of a growing number of NGOs, millions of hectares of heavily degraded farmland became wooded cropland. As with many other gm/cc systems, FMNR has become a platform upon which farmers can develop a multitude of new activities. In the process, farmers have increased the environmental quality, biodiversity, net income and productivity of their farms.

In Central America, we have strategically chosen to start working with the systems that could have the greatest potential impact around the developing world. These were, of course, those systems that included the major subsistence crops: maize, rice, sorghum, millet and cassava. In time, as good systems have been developed and spread for these large-scale cropping systems, we have branched out to working with systems dealing with vegetables, tree crops, and those designed for drought-prone areas.

We do not have proven solutions for three important agricultural niches. The first is that of paddy rice systems in which three rice crops are grown each year. To solve this problem, we are searching for leguminous plants that grow in swamps. A few months before this book went to the editors, we received word from I Made Setiawan, a young Indonesian agronomist, of an edible, spongy-stem crotalaria that has been traditionally used as a gm/cc for paddy rice in waterlogged areas across the Indonesian island of Kalimantan. This could turn out to be the solution for maintaining soil fertility in intensive paddy rice systems. (See S69)

The second niche is that of certain high-value, irrigated farming systems. Here the challenge is not so much a lack of possible gm/ccs, but the fact that the economics of these better-paid systems mean that compost and synthetic fertilizers become economically feasible. In the case of irrigated vegetables, the land is so valuable that the spaces that might be occupied by gm/ccs are more profitably used for other crops. That is, the opportunity cost of the space the gm/ccs would occupy is too high. For now, dispersed shade and crawling species like crawling cowpeas, which take up virtually no usable space, are the only feasible options in these situations.

A third problematic niche is that of farmers who are producing the equivalent of over 4 t/ha of maize. Identifying and developing gm/cc systems for this situation has not been a priority for the NGOs that work with gm/ccs. Very simply, most farmers who have achieved this level of productivity are no longer poor.

Still, in spite of all the good work carried out by EPAGRI, IAPAR, World Vision/Australia and World Neighbors, if it were not for all those systems discovered by smallholder farmers, we would not have a worldwide gm/cc movement today. These thousands of anonymous farmer-innovators are the true heroes in the story of the modern gm/cc movement.

Chapter 3. Green Manure/Cover Crops Around the World

As millions of farmers have found, gm/ccs enjoy a surprisingly large number of advantages. Gm/cc systems therefore often exist in places we do not expect to find them. In fact, they may exist even where local agronomists say there is no such thing (see photo 6).

The easiest way to find existing gm/cc systems is to pay a short visit to the bean section of the local market, especially at the end of the rainy season. Often you will find seeds of at least three or four species recommended in this book. Of course, some are merely being produced as a market crop, but in many cases, some of these species are planted in part because they help maintain soil fertility. By asking the merchants where they buy these species, you can often find out where farmers are using gm/ccs.

Photo 6. Indonesia. A number of Indonesian agronomists assured me that gm/ccs were not grown in their country because Indonesians do not eat beans. But our first lunch in a roadside cafe on Sumatra showed that they do eat tofu, string beans and mung bean sprouts, all made from locally-produced gm/ccs.

When asking merchants about prices, we must take some of their figures with a grain of salt (just as we must do with some of the information given by villagers). Small-scale merchants almost never give out accurate prices, especially if they know you may want to buy or sell some product they handle, in part because they are expecting to have to bargain before arriving at the going price.

The Advantages of Green Manure/Cover Crops

Gm/ccs' long list of advantages includes:

1. Increased SOM

Scientists hotly debate the amount of increase in SOM that can be achieved with gm/ccs over the years. In part this disagreement has occurred because some systems produce several times more OM than others. Factors causing this variation include differences in rainfall, temperature, gm/cc management, the number of gm/cc species in a field and the amount of biomass they each produce.

Photo 7. Honduras. The material in this photo may look like almost-finished compost, but it is, in fact, the soil from a typical field devoted to the 3rd maize/mucuna system (S52), after having produced maize every year for 40 years. Some people argue that soil fertility improvement over time is impossible without the use of synthetic fertilizer.[6] This experience, like that of literally millions of farmers around the world, proves that idea false.

The leaves in the background are those of the mucuna. They have been perforated by insects, but this merely means that some of the OM has been processed more quickly than usual. Insect attacks on gm/ccs only need worry us if the attacks are so heavy that they significantly slow down the vigorous growth of the gm/cc, or they affect the legumes' production of edible plant parts.

Some gm/ccs can add well over 50 t/ha of OM (green weight) to the soil each growing season (photo 7). Time and time again, I have found that in a field with 5 to 10 years of gm/cc use, the dark-colored, spongy soil, or topsoil, is 15 to 25 cm thick, yet 3 m away in a field where no gm/cc has been used, the topsoil is less than 5 cm thick. I have walked across fields like those of the 3rd maize/mucuna system (S52, photo 47) with a 40-year history of gm/cc use. It feels like walking on a soft mattress. If you dig down into the soil, you will pull up as much as 40 cm of what appears to be compost (photo 7).

Many people find it hard to believe that topsoil can be created so quickly. Many of us have heard the old cliché that it takes 100 years to build an inch of topsoil. However, this statement refers to nature's ability to create topsoil out of pure rocks. Few farmers these days need to start out with a pile of rocks. In fact, most farmers using good gm/cc systems are building up topsoil by about 2 cm/year, or more than one inch every two years. This may sound preposterous, but a little arithmetic can substantiate that fact.

A good gm/cc can produce from 40 to 60 t/ha of biomass (green weight) each year, which translates into roughly 10 t/ha dry weight, or 10,000 kg/ha. Since 1 ha contains 10,000 sq m, that translates into 1 kg/sq m. of dry biomass. A cubic m of soil averages about 1,760 kg in weight. Two cm of new topsoil in each sq m of land would be 2% of those 1,760 kgs, or about 35 kg/sq m of land. If we define topsoil as soil having at least 3% OM, 3% of 35 kg comes very close to the 1 kg of OM/sq m we have added to the soil. Therefore, mathematically, it becomes quite clear that our 40 to 60 t/ha of biomass can provide enough OM to turn 2 cm of totally inorganic soil into a moderately rich, 3%-OM topsoil. Since even badly depleted soils usually have between 0.5 and 1% OM before the gm/ccs are used, our new 2 cm of topsoil/year would more likely have 3.5 to 4% OM.

Applications of OM to soil provide an impressive number of positive effects on the soil: nutrients are recycled, phosphorus (P) and other nutrients become much more available to plants (especially in highly acidic or alkaline soils), rainwater infiltration increases over time to as much as three times its previous level, the soil's water-holding capacity increases and erosion is reduced. SOM can also improve the soil's nutrient balance, dramatically increase the number of macro- and microorganisms (which also help crops grow better), improve or "buffer" the pH of the soil and sequester atmospheric carbon. SOM softens the soil, making it resemble a sponge, rather than a concrete block, so air, roots and water can more easily penetrate into the soil. Plentiful SOM can also double or triple the formation of the nodules that enable gm/ccs to fix N.[7]

While serving as a mulch on the soil surface, OM can help maintain soil moisture content, control weeds and protect the soil from excessive heat. Nutrients in the mulch become accessible to plants and allow crop roots to access those nutrients before they become tied up in the soil. A good layer of mulch intercepts raindrops so they do not batter the soil directly, breaking up soil particles that thereby become more susceptible to erosion.

SOM also provides the food to support a major increase in the biological activity in the soil. The use of gm/ccs can multiply the number of earthworms/ha by 10 to 20 times, while also dramatically increasing the soil's microbial life. More and more evidence, from both the field and the laboratory, shows that an increase in biological activity, especially of mycorrhizae and other fungi, raises crop productivity substantially. Amazingly, recent research suggests that soil microorganisms can move up inside plants and enable them to grow faster, resist diseases better, and even develop more resilience to drought.[8]

2. Nitrogen fixation

In partnership with micro-organisms called rhizobia, legumes "fix" nitrogen (N), taking N out of the air and converting it into forms that plants can use (photo 8). This fixation of N is very important to smallholder farmers, because almost universally, N is the limiting nutrient in their soils.[9] This fact means that an increase in N will, at least initially, increase yields more than an increase in any other nutrient. That is, for most of the world's smallholder farmers, N is the only nutrient that will initially make possible a significant increase in yields.

Photo 8 by Laureano Jacobo. Honduras. Our textbooks usually show photos of N-fixing nodules as tiny spheres of about 1 mm in diameter, well-spaced along a root. Clearly, tropical legumes that fix from 80 to 400 kg/ha of N would produce more N than could fit in such minuscule nodules.

This photo shows the nodules from one single mucuna plant grown under ideal conditions with no competition from other plants. Under normal field conditions, a single plant may produce up to one or two of these "golf balls" of nodules per plant.

I know these nodules were from one single mucuna plant because even I did not believe what the farmers were telling me, so I asked to watch them dig up the nodules. I am the person holding the nodules in this photo.

Legumes, through N fixation, can add large quantities of available N to farmers' soils. The most widely used gm/ccs fix from 60 to 240 kg N/ha, while a few gm/cc species can fix up to 400 kg/ha of N. The exact amount of N that any species fixes varies tremendously from one place to another. Charts will list specific figures for specific amounts for different species, such as 80 kg N/ha for the *Vignas* or 150 kg N/ha for lablab beans. The National Academy of Sciences, for instance, lists mucuna as fixing up to 140 kg N/ha, the jack bean up to 240 kg N/ha and tarwi (*Lupinus mutabilis*) and fava beans up to 400 kg N/ha.[10]

Nevertheless, the N actually fixed by a particular legume in a specific setting will depend on a whole series of factors. If it is intercropped or planted at too low a density, it will often produce half of what it could potentially produce, or even less. If the soil has a pH of less than 5.0, it will produce a good deal less N than that listed. If synthetic fertilizer is used, or the soil already contains a good deal of N, it will also fix a lot less. Poor growing conditions of any kind, or a lack of rhizobia in the soil, can reduce N fixation down to almost nothing.

Still, other things being equal, yields will generally increase noticeably as we switch from common beans (up to 50 kg N/ha) to *Vignas* (about 80 kg N/ha), to mucuna and lablab beans (about 140 kg N/ha), and on up to jack beans, fava beans and tarwi (240 to 400 kg N/ha).

In most cases in developing countries, we do not have to worry whether or not a particular species is fixing maximum levels of N. For even maize crops that are producing 3 t/ha of grain, 100 kg/ha of fixed N should provide more than enough N. The one exception to this rule would be if we are using *Vignas* or some other legume that fixes relatively small amounts of N where farmers are harvesting more than 2 t/ha of maize.

A very common case of this problem is that of common beans (*Phaseolus vulgaris*). These beans usually fix less than 50 kg N/ha, and are therefore relatively useless as gm/ccs where N is the limiting factor in production. For that reason, this book does not generally include common beans as a useful gm/cc.

It is important that we appreciate the monetary value of N fixation. Even just the 140 kg N/ha that mucuna adds to the soil is roughly equal to the N in six 100-lb sacks of urea, which would cost most farmers in the developing world at least US $ 150.00/ha if purchased as a synthetic fertilizer.

The increase in both N and OM brought about by gm/ccs can double and triple soil productivity over time. As a result, programs from India to Brazil and Central America now use terms like "soil restoration" or "soil recuperation."

3. Astonishing resistance to drought

Probably the greatest surprise we have experienced in working with gm/ccs has been the amazing ability of these plants to provide resilience to drought. We expected improved fertility all along, in part because forests have managed, through traditional fallows, to keep the tropical world's soils fertile for millennia. But we somehow had not taken into account that forests untouched by people are also highly resilient to drought.

During the last five years I have worked mostly in the drought-prone lowlands of Africa. Droughts have been hitting many of these areas about every other year. These droughts, caused more by low rainwater infiltration rates than by a lack of rain, have destroyed most of the farmers' crops. Yet gm/ccs' ability, over time, to help crops do very well—even crops like maize, which are fairly prone to damage by drought—has amazed everyone involved. Chapter 5 describes the nature and reasons for this ability of gm/ccs to create resilience to droughts.

4. Resilience to climate change

The two most important impacts that climate change is having on agriculture are increasing temperatures and more irregular rains. However, dispersed trees can reduce the temperature of people's crops by up to 10°C., which will take care of the increased temperatures expected to take place over the next couple of centuries (see the section on "Dispersed or Intermittent Shade" in Chapter 10). At the same time, the increased resistance to drought can provide resilience to irregular rains. Therefore, gm/ccs have the ability to largely take care of the impact of climate change on agriculture for at least the next hundred years.

5. Carbon sequestration

Carbon dioxide is the primary cause of climate change. Carbon sequestration refers to the long-term removal of carbon dioxide from the air by storing it somewhere else. Since the soil, for millennia, was the world's principal carbon sink, by putting atmospheric carbon back into the soil, gm/ccs are merely helping humanity return the world to its previous natural state.

Gm/ccs can play a significant role in soil carbon sequestration both because they drop huge amounts of carbon onto the soil, some of which eventually becomes humus, and because they can be used on vast areas of the planet. Most good-quality gm/cc systems produce 10 to 20 t/ha/year of above-ground biomass (dry weight), though this can vary widely, depending on the system and its environment. The percentage of soil carbon that achieves long-term stability in the various conditions of tropical soils is still not well-known, although it tends to vary from 20% in poorer soils to 70% in soils rich in OM.[11]

If we take as an example a good gm/cc system using pigeon peas intercropped in maize with a crawling cowpea underneath, smallholder farmers occupying 12 % of the world's agricultural land could sequester 10% of all the carbon that needs to be drawn out of the atmosphere by the year 2100 according to the Paris Accords.[12]

Gm/ccs reduce atmospheric carbon in another major way. They reduce the use of synthetic fertilizer, which requires the burning of tremendous amounts of fossil fuels, both in their manufacture and transport.

By both creating major resilience to climate change and acting as an important carbon sink, gm/ccs also constitute the least expensive and simplest form of what is called "climate-smart agriculture."

6. Zero transportation costs

Unlike most other materials used to improve the soil, including synthetic fertilizer, compost and animal manure, the additional OM and N from gm/ccs incurs no transportation costs. They are produced in the very fields where they will be used and are already well-distributed across those fields.

7. Very low, or negative, net cost

If a gm/cc does not produce food or some other valuable commodity, it will not be adopted unless it costs the farmer an absolute minimum of cash. Hopefully, the first handful of seed is the only cash cost. If a gm/cc is attacked by insect pests, we generally do not worry about it. A few perforated leaves are no problem at all; the insects are merely helping break down part of the gm/cc's biomass. However, if the attack is serious enough to appreciably slow the growth of the plant, or destroy enough of the seeds to endanger the replanting the following season, the farmer will have to do something. One possibility is to find a simple method of insect control using locally available products, such as sticky traps, alternative hosts, a predator, smelly plants or liquids, etc. If no such simple remedy works, the farmer may have to replace the gm/cc with another species that is more advantageous financially.

On the other hand, if a gm/cc produces food or some other source of profit, a moderate amount of expense can be justified. Normally, though, food-producing gm/ccs enjoy a "negative cost." That is, the grain legumes they produce are usually worth more than the costs incurred in producing them. Furthermore, they may produce fodder, firewood or some other benefit, such as reduced weeding costs. In other words, farmers using good gm/cc systems usually make money from the process of fertilizing their fields with gm/ccs.[13] It's as if the salesman at the fertilizer store was giving out free fertilizer, and then paid the farmers a bonus for taking it off his hands.

If any gm/cc needs an inoculant, farmers can usually find a natural way to increase the rhizobia, such as letting the gm/cc grow, even though poorly, for one year so the population of rhizobia will increase. Or, once again, they can look for a different species of gm/cc.

All this means that for most field crops, synthetic fertilizers just cannot compete on cost with gm/ccs for most smallholder farmers.

8. Weed control

Gm/ccs can also dramatically reduce the cost and labor required for controlling weeds (photo 9). Especially in Africa, this often means that the time women spend weeding can be reduced by anywhere from 20 to 90%. In southern Paraguay, the cassava/jack bean system (S75) eliminates two of the three weedings normally needed to grow cassava (i.e. about 65% of total weeding labor). In the 6th maize/mucuna system (S55), farmers produce two crops/year

Photo 9. Honduras. Aside from four or five blades of grass, there are no weeds in this photo. The mucuna in the photo is the most efficient known gm/cc for controlling all kinds of weeds, but lablab beans, jack beans, perennial peanuts (*Arachis pintoi*) and runner beans can do nearly as well.

If we calculate the savings achieved by farmers who only need to weed their crops once a year, we find that farmers using these gm/ccs often fertilize their fields well while significantly *reducing* their total costs, including labor.

of maize, but only need to weed the first crop; the mucuna mulch from the first crop totally eliminates the need to weed the second crop (photo 48).

In the 3rd maize/mucuna system (S52 in photo 47), things get even better. Farmers using this system have planted maize for up to 40 years, yet after the first two or three years spent establishing the system, they have never had to weed their fields again, except in occasional very small patches where the mucuna somehow failed to reseed. This system saves at least a 95% of people's weeding labor.

In Africa, a particularly noxious parasitic weed is striga (*Striga hermonthica*). Striga can significantly reduce yields of maize, sorghum and millet. The solution, however, is quite simple. Striga only grows in depleted soils. Therefore, it will die out if enough OM is added to the soil. Almost any gm/cc that produces an above-average amount of biomass will rid a field of this weed, often in three to four years.[14] Controlled experiments in Kenya have shown that the use of legumes can reduce striga counts by 68% in just two years.[15]

Another serious weed, Imperata grass (*Imperata cylindrica*, often called "speargrass" in Africa and "cogon grass" in parts of Asia) also grows best in poor soils and therefore becomes less problematic as soils are restored. Three or four years of gm/cc use can largely eliminate it.

9. Reduction in the use of agricultural chemicals

If farmers are already using synthetic fertilizer, gm/ccs can reduce the amount used by up to 60% without lowering yields. Herbicide use can be reduced or eliminated. Some species of gm/cc can be used in place of other chemicals. For example, mucuna, lablab beans and many other gm/ccs can control nematodes,[16] while sunn hemp (*Crotalaria juncea*) leaves can control the weevils in stored grain.

10. Soil cover and erosion prevention

The soil cover provided by gm/ccs can be very important for soil conservation. In general, but especially in the tropics, the value of soil cover has been greatly underestimated. High rainfall and steep slopes without cover can lead to ruinous levels of erosion. Raindrops falling on bare ground break up the soil into smaller particles that are then easily washed down a hillside. Covered soil is much less vulnerable.

One study in Brazil shows that well-managed gm/cc systems using zero tillage "can reduce erosion to less than 5% of that observed under conventional soil tillage."[17] A study in Kenya shows that a cover of mucuna reduced erosion to 17% of that on bare plots.[18] A third study reported that farmers using the 3rd maize/mucuna system (S52) on hillsides with a 30% slope and more than 2,000 mm of annual rainfall were actually increasing the productivity of their soil year by year, with no other soil conservation practices being used. This "negative erosion" is occurring because the soils are not plowed, but rather are covered by the mucuna or its mulch 12 months of the year.[19] The mucuna has also been building up the OM content of the soil for decades, thereby increasing the soil's infiltration rate and water-holding capacity.

Soil cover is provided by green mulches (when gm/ccs are growing) and dead mulches (when they have died and are lying on the soil). Fertilizer trees, used either as dispersed or intermittent shade, also cover the soil and reduce the impact of raindrops.

11. The capture of wind-blown soil

In regions like the African Sahel, wind erosion is a major problem during the month of dry season winds known as the "harmattan." Low-lying bushes that grow leaves year-round, such as *Piliostigma* (*Piliostigma spp*), can capture soil that is blowing in the wind, adding important nutrients to farmers' fields. Over several years, up to 40-cm-tall piles of topsoil can gather under each bush.

12. Zero tillage

The experience of over 3 million farmers in Brazil, Paraguay, Argentina and Honduras shows that using good gm/ccs, farmers can adopt zero-till systems that continue to produce 4 t/ha of maize or more, with no "transition period" of lower productivity. Farmers using gm/cc systems that include rotations and moderate applications of synthetic fertilizer regularly harvest 7 to 8 t/ha of maize without having tilled the soil in a decade or two.[20] By adding gm/ccs to zero-till Conservation Agriculture, African farmers suffer no reduction in productivity during the transition and can triple their yields within five to seven years, while reducing tillage costs.

In Honduras, Guatemala and Mexico, using mucuna, zero tillage and no synthetic fertilizer, smallholder farmers using the 3rd maize/mucuna system (S52) have maintained yields of maize at over 3 t/ha for up to 40 years. Their maize yields increase to 4 t/ha with a very small additional application of urea.[21]

Thus gm/ccs can help farmers transition to zero-till farming systems that save them a tremendous amount of time, effort and expense because they no longer need to plow the soil. Improving soil health and productivity provide an added bonus.

13. Competitiveness with farmers using tractors

Since tillage is the primary method used in conventional agriculture to control weeds and prepare the soil, wealthy farmers who use tractors to speed up these operations have always enjoyed a huge advantage over smallholder farmers. Gm/ccs, however, can smother weeds and facilitate zero tillage. Thus, smallholder farmers using animal traction, or even hoes, plus gm/ccs, can now compete with tractor owners. After all, eliminating these operations altogether costs a lot less than just mechanizing them. In an age of falling trade barriers, this advantage alone could justify the widespread use of gm/ccs by smallholder farmers.

A very clear example of this situation is provided by the 3rd maize/mucuna system (S52). With this system, smallholder farmers' costs of producing each sack of maize is about 30% less than that of the richer conventional farmers nearby who use tractors, synthetic fertilizers and herbicides on flat bottomland. Since the yields of the maize/mucuna system are about 25% lower, the maize/mucuna system earns roughly the same net income per hectare on what was originally much less fertile land. Furthermore, to do so, they invest much less money, take fewer risks and increase their soil's natural fertility year after year rather than degrading it.[22]

14. The ecological advantages of trees

Farmers who use trees as gm/ccs, such as with dispersed or intermittent shade, also gain the other ecological advantages provided by trees. The trees protect the soil and crops from the tropical sun, conserve soil moisture by reducing evaporation rates, transpiration rates and wind speeds, curb desertification, sequester carbon in their trunks and branches, and pump soil moisture and nutrients up from deep in the soil. Just the reduced heat under a tree's moderate shade may increase crop yields from 20 to 30%. The presence of trees makes a field become more and more similar to a forest, and forests have proven they can maintain their soil's natural fertility and environmental health for tens of millions of years.

15. Resistance to insect pests and plant diseases

Two different aspects of gm/cc systems can make our crops much less susceptible to pests and plant diseases. The first is intercropping. A study carried out by the Zamorano Panamerican Agricultural University in Honduras showed that intercropping mucuna with maize brought a reduction in two of the most important insect pests and diseases of maize.[23] On the downside, it caused a small increase in a third pest and a third disease.

The corn borer worm is probably the worst insect pest of maize in Honduras. Its population in the intercropped experimental plots was half of that in the monocropped control plots. This result apparently occurred because when the adult females flew over the intercropped field looking for a maize field in which to lay their eggs, the presence of the mucuna reduced their ability to smell the maize. About half of them therefore flew on to other plots. The increased biodiversity of an intercropped plot often confuses insect pests in other ways, or provides more hosts for their predators. In Uganda, for instance, the presence of intercropped soybeans or peanuts reduced the termite damage to maize because the legumes attracted a lot of ants—the termites' worst enemies.[24]

The second way that gm/ccs help control some insect pests and plant diseases is through the presence of increased SOM levels. For instance, Brazilian scientists discovered in the 1980s that mucuna controls nematodes. Then we discovered in Honduras that lablab did the same. More and more, we are finding evidence that large amounts of SOM, regardless of its source, depress most species of nematodes. How does SOM control nematodes? Some researchers say that SOM provides a wide variety of sizes of pores in the soil, providing more hiding places for the nematodes' predators. Others add that large quantities of rotting SOM produce chemicals that kill certain pests and diseases.

Some gm/ccs contain specific chemicals or toxins that help control certain pests. For instance, the natural fungicides in jack bean leaves can kill the fungi that leaf-cutter ants grow for food. If a farmer plants jack beans near a leaf-cutter anthill and the ants carry jack bean leaves down into their burrow, the fungi will die and the ants will either starve or move elsewhere.

16. Inexpensive recuperation of wastelands

Farmers can use several different gm/cc species to recuperate wastelands, but jack beans and pigeon peas are usually the best. Jack beans are more capable of growing in highly depleted soils and of withstanding droughts than almost any other gm/cc species during their first year of growth. It can fix more N/ha than that in 11 100-lb bags of urea. By plowing a wasteland and planting jack beans at 2 to 3 seeds/sq m at the beginning of at least a 5-month rainy season, we have been able to produce an 800 kg/ha maize crop the following year. The only three cases in which I have seen the jack bean grow too poorly to restore the land were in central Cambodia, in Rwanda, and on the Kalahari sands of western Zambia. In all three cases, the soil had a pH below 4.5.

17. Long-term yield increases

The main motivation for using gm/ccs is, of course, to achieve long-term increases in crop yields. Probably the best yield reported from formal experimentation with maize yields following a gm/cc was from Brazil, where maize following mucuna, without synthetic fertilizer, produced 7.4 t/ha.[25] Experiments in Kenya showed that on average, "legume treatments increased maize yields by 120% over the no-fertilizer control (1.05 t/ha)." Mulching with mucuna "increased maize yield from the no-fertilizer control by 300%," at a 1% level of confidence.[26]

Widespread experience with gm/ccs in farmers' fields that produce less than 1.3 t/ha of basic grains without fertilizer, indicates that any good gm/cc system should increase maize yields by at least 50% at the end of the second season (that is, the first season after the biomass is applied), and by at least 100% within three to five years.[27] If farmers are not achieving at least these levels of increased yields, they may be doing something wrong. The most frequent mistakes include those of using the wrong species of gm/cc, using only one such species or experiencing problems with volatilization or synchronization.

18. Soil restoration is a "foundational technology."

The restoration of soils is what we call a "foundational technology." If you introduce a new variety of maize, it is only going to help farmers produce more maize. On the other hand, if you teach them to restore their soil (or manage their water resources better or improve other important environmental conditions), you can help improve absolutely everything they might do on their farms. In modern terminology, the introduction of foundational technologies can create a "platform" that enables farmers to begin experimenting with all sorts of new possibilities. They may become able to grow a number of new crops or animals of better quality than ever before, thereby increasing their incomes, increasing their biomass production, improving the biodiversity of their farms and ultimately producing more food for hungry nations.

Such foundational technologies, together with famers' learning how to experiment and share knowledge with each other, can often, over time, provide the basis for achieving a whole new quality of life.

19. Additional important benefits

In addition to the above advantages, specific gm/cc species can provide one or more of several additional benefits:

- high-protein human food;
- production of a wide range of commercial products;
- high-protein fodder and bedding for domestic animals;
- contour hedgerows for erosion control; and
- firewood.

Of course, any time we use gm/ccs as food, fodder or fuel, we reduce the quantities of N and OM added to the soil, thereby reducing the gm/ccs' ability to improve soil fertility and increase resistance to droughts.

Each of the above advantages should be analyzed carefully when selecting a gm/cc system. Experience in a score of countries shows that, initially, farmers are rarely attracted primarily by the gm/ccs' ability to increase soil fertility. More commonly, farmers are motivated mostly by the gm/ccs' potential for producing human food or controlling weeds. Therefore, gm/ccs should be promoted by emphasizing their other advantages as much as that of soil fertility. For instance, when talking with farmers, we sometimes refer to a gm/cc as a "food crop" or "green herbicide," rather than a gm/cc.

As farmers acquire more experience with gm/ccs, they come to believe in and appreciate more all their advantages, including the long-term ones, so they are happy, if need be, to include additional gm/cc species that provide fewer supplementary benefits.

The Disadvantages of Green Manure/Cover Crops

Despite all these advantages, we commonly encounter difficulties when introducing gm/ccs to smallholder farmers. During the 1980s, I personally tried for nearly eight years to introduce mucuna to farmers in central Honduras, with very little success. This occurred because at the time I did not understand some of the important disadvantages of gm/ccs. If we do not acknowledge and overcome these disadvantages, our efforts to introduce gm/ccs will likely fail.

1. Space for planting gm/ccs

Farmers will normally refuse to plant a non-food-producing gm/cc, and even many food-producing ones, on land when and where they can produce either food or a cash crop. That is, the space (including the space between other crop plants) or time of year at which one grows the gm/cc must have no other valuable locally known use. For example, if farmers already grow common beans as an intercrop in all their maize fields, they will normally refuse to intercrop an inedible gm/cc there. But if farmers are monocropping at least some of their maize, leave their land bare during the dry season, or still fallow their land, possibilities for incorporating gm/ccs will likely abound.

2. Slow results

Soil improvement is unfortunately a slow process in which farmers probably will not notice any positive results immediately. Significant yield improvements usually occur only after a gm/cc has been applied to the soil, so visible results do not normally appear until half-way into the second cropping cycle. Furthermore, the results are often hard to believe. After all, smallholder farmers have watched their soils deteriorate year after year, but unless they can remember their fathers' or grandfathers' long-term fallows, they have rarely ever seen a soil improve.

Once again, it may help to promote gm/ccs for reasons other than soil fertility. If the gm/ccs produce additional foods that farmers are familiar with, farmers may see recognizable results within a month or two. The same thing can happen if farmers' primary motivation for planting a gm/cc is to achieve better weed control.

We can also show farmers gm/ccs' impact on fertility in simple, more immediate ways. For example, we can bury a double handful of fresh gm/cc leaves at the base of each recently planted vegetable or maize plant in one part of a field. The results of such a simple experiment can go a long way toward motivating farmers to try a gm/cc. Simple school-ground demonstrations can also illustrate how OM increases the water-holding capacity of soil, or how mulching can reduce erosion. In drought-prone areas, we can demonstrate how OM increases rainwater infiltration by pushing a 40-cm long piece of plastic pipe 15 cm into a soil high in OM and another one into heavily depleted soil. We then pour an equal amount of water into each pipe, timing how long it takes in each case for the water to disappear into the soil.

We are just beginning to develop some gm/cc systems that produce much faster results. The 60-day cowpea is an important example. If planted during the first rains, before the rainy season has really begun, these cowpeas can be harvested and the crop residues applied to the soil as early as a month into the rainy season, thereby acting as a side dressing of N for the main crop during the first year of the cowpea's use. The visible impact of the cowpea thereby occurs early in the same season in which it was planted.

3. Dry season problems

During the dry season, wild animals, free-grazing domestic animals, termites, agricultural burning or bush fires may destroy the OM provided by a gm/cc. If domestic grazing animals or wild animals are causing the problem, we need to either build or grow fences (such as the monkey thorn fence) or use gm/cc species that the animals will not bother, such as tephrosia or jack beans.

In the lowland tropics, with five to eight months of dry season and no dispersed shade, the tropical sun burns off 60 to 80% of the N of a gm/cc's crop residues in a process called volatilization. As much as 30 t/ha/month (green weight) of its biomass will be burned off, as well.[28] The following crop will therefore benefit very little from the gm/cc. In such cases, we must either grow adequate shade to reduce the ambient temperature and therefore the volatilization, or use gm/cc species like jack beans, pigeon peas, lablab beans, tephrosia or *Piliostigma*, that continue to grow through most or all of the dry season.

Another solution, which can protect the fertility of gm/ccs' leaves from both grazing animals and volatilization, is to grow fertilizer trees as dispersed or intermittent shade, with their high-N leaves growing through most or all of the dry season and well out of reach of the animals.

4. Difficult growing conditions

Smallholder farmers have to cope with many challenging conditions, including extremely low or irregular rainfall, heavily depleted soils, extremes in soil pH, severe drainage problems, flooding, or a combination of these. Such conditions slow the growth of gm/ccs, reducing their impact.

Through the years, we have learned how to overcome many of these problems. We can often solve them by using gm/cc species that are particularly suited to the negative conditions involved. For example, jack bean can withstand very poor soils and serious droughts. Ratooned or pollarded gm/ccs (i.e. on which most of the stalk has been cut off) like the ratooned pigeon pea or gliricidia (*Gliricidia sepium*), withstand drought extremely well.

However, gm/cc species that work well under such conditions, like the jack bean or tephrosia, often lack major additional benefits or fit poorly into the local farming system. In some cases, such as that of depleted soils or irregular rainfall, farmers may accept planting one of these less useful species for only one or two years until the soil is more fertile and drought-resistant, and then switch to a gm/cc species that provides more additional benefits.

An additional expense in the case of using gm/ccs for restoring wastelands is that many of the viny- and bushy-type legumes will not grow well on an established grassland. For these species to flourish, the land must first be used as a cropped field, fallowed with trees (so that grassy weeds are at a minimum), or plowed.

5. Timing

To raise productivity, the nutrients provided by a gm/cc must become available to crops precisely when the crops most need them. A simple rule of thumb is to have lots of green leaves on the ground when we plant our crops. With many gm/cc species, achieving the correct timing (or "synchronization") is difficult, causing yields to suffer. We can sometimes solve this problem by finding a gm/cc species that grows longer into the dry season, or by relaying a shorter-lived gm/cc into people's crops. In many cases, a dispersed shade of gliricidia or *Piliostigma*, in addition to the original gm/cc, might well solve the problem by reducing N volatilization. If none of these solutions works, we might use natural foliar nutrient sprays, such as solutions made from cattle urine or of crushed gliricidia leaves, to supplement the needed nutrients at the right time.

Frankly, much of the challenge of selecting the best gm/cc system for any given area or farming system consists of trying to minimize these disadvantages.

The Characteristics Desired in a Gm/cc Species

The characteristics farmers desire most when looking for a gm/cc species include its ability to:

1) Grow vigorously, producing tons of biomass as quickly as possible.

2) Fix as much N as possible, usually a minimum 60 kg of N/ha/year.

3) Grow well in poor conditions, including drought, poor soils or heavy shade.

4) Require little additional labor and no fertilizer or expensive inputs.

5) Provide important benefits beyond those of fertilizing the soil, maintaining soil moisture and controlling weeds. Farmers rarely become enthusiastic about a gm/cc that provides no benefits beyond fertilizing the soil.

Though these characteristics can be briefly stated, they can be tremendously important in achieving success with gm/ccs.

Why Not Just Use Synthetic Fertilizer?

Some people might ask why we do not just use synthetic fertilizer. Quite simply, synthetic fertilizer has become unprofitable for most smallholder farmers producing basic subsistence crops like maize, sorghum, millet and cassava.[29] The situation differs for each individual farmer and each crop, but in most cases the cost of using synthetic fertilizer is now greater than the value of the increased harvests it provides (photo 10). A few major exceptions to this rule still exist, the most important being that of rice. Rice enjoys a relatively high price on the world market, making it capable of paying many farmers for their use of synthetic fertilizer. Farmers who grow high-value vegetables and fruit can also often make profitable use of synthetic fertilizer.

Photo 10. Haiti. Synthetic fertilizer has a whole series of "hidden" costs that rarely show up in calculations of its profitability. With gm/ccs, the costs are dramatically reduced.

The Haitian farmer in this photo must rent or own an animal and transport the fertilizer for 6 hours from the road head to his home. He would never have bought this fertilizer if its cost had not been subsidized by a non-governmental organization (NGO). Subsidizing fertilizer creates an unsustainable situation; farmers often stop using it the minute the subsidy ends. Subsidization of fertilizer also acts as a major disincentive for farmers to learn about more sustainable technologies.

Some scientists still publish calculations purporting to show that fertilizers are profitable for smallholder farmers. But if all the fertilizers' costs are included, farmers usually end up losing money. These costs include taking a bus to a local town; purchasing the fertilizer at a small-town retail price; returning on another bus and paying for the transport of the fertilizer; and the labor required to apply the fertilizer. We must also add in the prohibitive cost, for smallholder farmers, of purchasing, transporting and applying lime to neutralize the acidification caused by the fertilizer (fertilizers containing N can destroy an acidic soil's productivity if the fertilizer's acidity is not neutralized periodically). If soils are acidic and lime is not available at all, farmers risk ruining their soils completely. And then one must add in the cost inherent in the increased erosion caused by fertilizer, and most important, the risk of losing the entire investment if the crop fails. In many drought-prone areas of Africa and Latin America, droughts now occur at least once every third year. Since they cannot be predicted, applying fertilizer at the beginning of the season in a drought-prone area involves a huge risk—more than enough to make synthetic fertilizer use on basic grains completely unprofitable.

To make matters worse, many of these published calculations quote outdated fertilizer prices far below today's prices. As a result, smallholder farmers are often criticized for refusing to use synthetic fertilizer, when, in fact, they may be quite correct in doing so.

But the real issue here is not whether synthetic fertilizers are profitable. If an alternative technology exists that is vastly cheaper, is widely accessible, and provides many other nutritional, economic and ecological advantages, we should not be recommending the more expensive option, even if it were profitable.

Of course, if we assume the area has good rainfall, lime is both locally available and cheap, and fertilizer is profitable, a heavy application of synthetic fertilizer helped by a good supply of OM, could double productivity from a base of 1 t/ha in one year, instead of two or three years. Results would come more quickly and be more impressive than with using gm/ccs. But one must question whether it is worth a program's time and expense to become involved in a major agricultural credit operation in order to achieve those somewhat quicker results. It would seem to be preferable to spend less money per farmer by avoiding the credit operation, thereby spreading the results to far more farmers per dollar spent. Furthermore, using gm/ccs means that farmers can, and do, spread the technology spontaneously from farmer to farmer and village to village. Any process heavily dependent on outside cash will cut later spontaneous spread to virtually nothing.

Smallholder farmers around the world now possess an average of around 0.75 ha of land/family. Even a moderate application of synthetic fertilizer on 0.75 ha will cost at least US $ 150.00, in addition to all the other costs mentioned above. It is virtually impossible for a smallholder farmer to feed his or her family on what is left from the harvest of 0.75 ha of basic grains, even when fertilized, if s/he must sell enough of it to pay $ 150.00 for fertilizer. On the other hand, gm/ccs can improve soil productivity at a small fraction of that cost, if anything. By saving this $ 150.00, farmers can feed their families adequately on much smaller pieces of land. Taking this route is imperative if most of the world's smallholder farmers are to achieve anything approaching food security on 0.75 ha.

Comparing Gm/ccs with Animal Manure and Compost

Animal manure is, of course, an extremely valuable and effective soil amendment. Farmers should use as much of it as they can, and should always be taught how to manage it well. Just keeping the manure covered while it is stored, by using a simple thatch roof or a piece of plastic can double or triple its impact.

However, the vast majority of smallholder farmers have nowhere near enough manure to maintain their soil's fertility. It takes at least seven well-fed cattle to produce enough manure to sustain the fertility of just 1 ha of land. Farmers with that many animals are no longer among the poorest.

Compost costs a good deal more in labor than do gm/ccs. It also fails to provide the additional benefits of gm/ccs, whether they be food, fodder, weed control or firewood. And compost can only decompose already available biomass. In contrast, gm/ccs provide *additional* badly needed biomass to those farming systems.

Like animal manure, compost also has a problem of supply. If everyone in most villages wanted to use enough compost to maintain their fields' fertility, they would run out of biomass long before they reached their goal. In the process, they would leave nothing to feed their grazing animals. Furthermore, the labor cost of making anywhere near enough compost would be prohibitive. In all my travels over half a century, the only smallholder farmer I ever met who has ever made and applied compost on even a 1/8-ha of basic grains at a rate of more than 20 tons/ha (fresh weight) is Conrado Zavala (photo 5). He did it for one small experimental plot. He never did it again.

The exceptions to these observations would be those farmers who are growing high-value crops and/or own less than ¼-ha of land. In these cases, compost can play a very important role.

Photo 11 by Laureano Jacobo. Honduras. Composting is rarely profitable if used on basic grains or root crops, other than rice. Farmers will often continue to make compost piles when a program is present, but soon after the program ends, the compost making will end, too.

Experts in soil biology have recently begun pointing out that animal manure, as well as compost that includes animal manure, contains vast amounts of microbes, especially fungi, that enhance plant growth in many ways. They maintain that farmers should continue to apply animal manure to their fields even if they have more than enough biomass from gm/ccs to improve their soil. For this purpose, relatively small quantities of manure or compost can have an appreciable impact because they act more like an inoculant than a fertilizer. To date, we have not had time to test such inoculation techniques under smallholder farmer conditions.

But even given this impact of animal manure and compost, *the use of gm/ccs is the only practical way that most smallholder farmers can maintain the organic matter levels, and therefore the natural fertility, of the vast majority of their soil.* This conclusion may seem to some people to be a bit of an exaggeration. Nevertheless, it is the same one reached by the Brazilians after 30 years of teaching farmers to use Conservation Agriculture on some 25 million hectares: "On small farms, the only practical and economic means to maintain and/or increase soil organic matter levels is to utilize a green manure/cover crop, or a combination of GMCCs, that has a high potential for biomass production in order to complement the contribution of crop residues."[30]

Chapter 4. Improving Soil Fertility and Soil Moisture

A Major Misconception

In her excellent classic, *The Ecological Management of the Soil*,[31] Ana Primavesi describes an experiment that challenges the very foundations of conventional thinking about soil fertility.[32] In this experiment, maize was grown in four hydroponic solutions. The first solution contained a normally recommended concentration of nutrients for maximum plant development, and it was replenished every 4 days. In the second, twice the normal concentration was used and replenished every 4 days. In the third, the normal solution was diluted 50 times and also replenished every 4 days. In the fourth case, the solution was the same as in the third case, but was replenished every two days.

Plant growth, surprisingly, was less in the second case than the first. In the third case, as would be expected, plant growth was reduced, by 28%. But the most surprising case was the fourth case: plant growth was slightly superior to that of the first treatment. That is, *nutrients supplied to the plants at only 2% of the normally recommended concentration for optimum plant growth, produced the same results as a normal solution, as long as the nutrients were replenished often enough.*

This astounding result sounds impossible. Millions of farmers in temperate climates know that if they do not apply the amount of fertilizers recommended, they will suffer lower yields. So what is happening here? The fact is that constant access to nutrients over time, even if they are very small quantities of nutrients, is far more important than the total quantity of nutrients in the soil. We speak of the "Nutrient Access Theory," born of Ana Primavesi's findings, as being a more accurate description of soil dynamics than the conventional "Nutrient Quantity Theory."

But if that is so, why do yields fall when large-scale farmers apply less than the conventionally recommended doses of nutrients?

Several factors can explain this mystery. First, and perhaps most important, large-scale farmers tend to apply fertilizer only a very few times during a whole year. Second, soils, especially those seriously short of OM, tie up huge amounts of those nutrients. In the case of P, over 98% of it is often tied up within days of application. To provide constant access to these nutrients for a whole year, many times the required concentration of nutrients has to be applied. Third, access to many of the nutrients applied is limited because with little OM in the soil, the soil is hard to penetrate, it forms hardpans and claypans, and its structure is poor. Fourth, with less buffering of the pH by SOM, its pH will be more extreme. Fifth, mycorrhiza in the soil extend the reach of crop roots, thereby helping them access more of the nutrients in the soil. But mycorrhiza populations are significantly reduced by a lack of SOM and the presence of nitrogenous fertilizers. Finally, in cold climates, nutrients from SOM tend to decompose much more slowly than in the tropics, making them less available.

A number of well-known phenomena contradict the Nutrient Quantity Theory. For instance, humid tropical forests produce some of the world's largest amounts of biomass/ha on some of the world's most nutrient-deficient soils, year after year. But how can they? Nutrient pumping cannot explain this apparent contradiction to conventional wisdom because the amount of nutrients pumped from deep in the soil is nowhere near what would be needed. Furthermore, tropical grasslands, with much shallower root structures, do much the same thing as the forests.

Proponents of the Nutrient Quantity Theory have claimed for decades that West Africa's geologically very old soils have few nutrients left, so good crops cannot be grown there without the heavy use of synthetic fertilizers. But we have produced very good yields there using OM that provides only a small fraction of the nutrients at any one time that the conventional wisdom would recommend, but keeps making them available over time. And even the fertilizer companies are now contradicting their own recommendations by promoting micro-dosing all across the Sahel. If large quantities of nutrients are required for decent crop growth, especially in the poor soils like those of the Sahel, how can micro-dosing be of much use?

Lastly, virtually none of what this book is all about could be true if huge quantities of nutrients were needed for plants to grow well. Gm/ccs are, by their very nature, slow-release fertilizers that almost never produce the concentrations of nutrients that are recommended by the fertilizer industry.

That they work at all provides major evidence that large concentrations of nutrients in the soil at any one time are unnecessary.

If the quantity of nutrients in the soil is not the most determining factor for plant growth, what is? It is precisely what the hydroponic experiment indicates: the long-term, constant *access* to nutrients, even if in very small quantities. More precisely, in order to produce maximum plant growth as cheaply as possible, what we need to do is to:

1) provide a constant supply of soil nutrients, even if they are supplied at very low concentrations;

2) supply these nutrients in a healthy balance for plant growth; and

3) maximize the access of crop roots to these nutrients, through good soil structure and plenty of mycorrhizae and other microorganisms, or even easier, through a biodiverse mulch.

How can we achieve maximum nutrient access? The easiest way is by using a thick, year-round mulch. OM, if from a biodiverse source, is a slow-release, well-balanced source of nutrients. It can very easily be replenished throughout the year by a combination of natural leaf fall and the gradual process of mineralization in a thick mulch, which can fill in the gaps in leaf fall. Of course, this all assumes that we have a large amount of biomass with good biodiversity in the field. These nutrients will then be highly accessible to plant roots if accessed from the organic mulch. If the nutrients must be accessed through the soil, the soil must have a good structure, plenty of OM and plenty of microorganisms, including mycorrhizae.

Once we understand the Nutrient Access Theory, all the above contradictions—the results of the hydroponic experiment, the high biomass production of tropical forests, the results achieved by slow-release fertilizers and micro-dosing, and the widely validated success of gm/ccs—become totally understandable.[33]

Some readers may feel that surely soils must have nutrients or qualities that plant roots require which would not be present in the absence of soil, even in a moist, biodiverse mulch. Nevertheless, hydroponics has proven that crop plants can grow perfectly well in just water plus the essential plant nutrients. Since biodiverse plant material will inevitably contain all the essential plant nutrients in more or less well-balanced amounts, it is obvious that there is nothing present in a hydroponic solution that would be missing in a moist, biodiverse mulch.

Can Gm/ccs Improve Soil Fertility Over the Long Run?

For over 30 years, a major controversy has been raging as to whether soil fertility can be increased over decades without using synthetic fertilizer. Those of us who have worked with or observed gm/ccs, have suspected since the 1960s that gm/ccs were capable of improving soils over the long haul. After all, traditional farmers all over the tropical world who fallowed their land would grow crops for two to five years, their unfertilized harvests dropping from about 1.5 t/ha to perhaps 0.5 t/ha of maize, or the equivalent for other crops. Then they would fallow the land for 15 years or so, and when they farmed the same land again, it was producing 1.5 t/ha again. If plants could not improve soil fertility over the long haul, how did the natural vegetation of virtually everywhere in the tropics triple the natural productivity of forest land over 15 years, even though farmers were taking nutrients off the land on a regular basis when they farmed it?

By the mid-1980s, those of us involved with gm/ccs began to realize that planted gm/ccs were perfectly capable of accomplishing the same thing as forests, both because of centuries-old farmer-generated gm/cc systems we observed, and because of our own experiments. By the late 1980s, we were publishing materials telling of these successes in a number of magazines and books. Peer-reviewed journals from Brazil have also reported well-designed scientific experiments that have shown that gm/ccs were increasing soil fertility significantly over the long haul.[34]

Furthermore, literally millions of smallholder farmers have developed and used gm/cc systems, and with them have increased or maintained their crop yields over decades. The most closely studied of these is the 3rd maize/mucuna system in Mexico, Guatemala and Honduras. (S52 and photo 47) In this case, it is clear to anyone who visits farmers using this system that for at least 20 years these farmers have maintained maize yields of above 3 t/ha, starting from a base, usually, of around 1 to 1.5 t/ha. Bernard Triomphe, in his Cornell University doctoral dissertation, reported that in the villages he studied in Honduras, maize yields had increased from about 1.8 t/ha to 3.3 t/ha. He also reported that the SOM content of these soils had increased over the first ten years of using the system and then leveled out. Soil N content also increased and then stabilized, while available soil P had stabilized over the same period.[35]

On the other hand, the fertilizer and agrochemical companies, along with a lot of researchers who depend on them for financial support, have claimed for the last 30 years, that it is impossible to increase soil fertility over the long haul with gm/ccs. Well-respected sources like the National Academy of Sciences (NAS) have, for half a century, been publishing such information.

Take, for instance, the chapter on "Nitrogen and Organic Matter" from the NAS book *Soils of the Humid Tropics*.[36] It contains dozens of completely erroneous statements, all understating what gm/ccs can accomplish. One quote claims that "when a part of the crop is harvested and removed from the land, equilibrium in organic matter can exist only if the nitrogen and organic matter inputs are supplemented from sources apart from the land area in question." The truth is that we not only can maintain, but significantly increase N and SOM by more than what has been lost, by using intercropped gm/ccs on the same land that is being cropped. Another quote claims that "green manure crops and annual legumes may provide from 10 to 70 kg/ha [of N]." In fact, the fixation of N from just one single annual legume can go as high 400 kg/ha/year. With multiple legumes (the possibility of which was not even considered by many researchers until the last few years) we can do even better. And the NAS book purports to be talking specifically about the tropics.

A more recent example from 1993 from a very different scientific source is the book *Soil Chemistry and its Applications* by Malcolm Cresser, *et al.* This book admits that at least some soil improvement is possible, but speaking of a best-case scenario: "To increase this [3% OM content] to 4% organic matter would take close to 100 years if 2 tons/ha [dry weight] of crop residues were added annually."[37] Once again, it was known by the early 1980s that with gm/ccs, 6 to 8 t/ha of OM could be added to the soil very cheaply from intercropped gm/ccs, increasing soil humus by at least 20 times the stated amount, while the soil was being farmed every year.

Lastly, ICRAF, in 2012, just four years before its scientists apparently dramatically changed their minds, according to the *Scientific American* article cited a few paragraphs below, was still proclaiming that gm/ccs did not work.[38]

Perhaps the most exhaustive book on the subject of long-term soil fertility improvement with tropical gm/ccs is Andre Bationo's *Lessons Learned from Long-term Soil Management Experiments in Africa*. It states that one of the "key lessons" learned from 19 long-term trials carried out across Africa is that "all long-term trials showed yield decline, often with a relatively rapid fall to a low level of equilibrium. Such yield dynamics are highly correlated with available nutrients and soil organic matter."[39] Bationo concluded this even though he was referring to trials some of which included applications of synthetic fertilizer!

What can explain the discrepancy between Bationo's conclusion and one of the main theses of *Restoring the Soil*—that long-term improvements in fertility can be achieved fairly easily? For one thing, some of the experiments reported in Bationo's book did not even use gm/ccs. More importantly, all but one of the experiments with gm/ccs violated the rules described above for achieving long-term soil fertility. They almost all used common beans, peanuts (*Arachis hypogea*), soybeans (*Glycine max*) or common cowpeas. The first three species seldom produce anywhere near enough biomass to raise yields. The common varieties of every one of these species dies three to four months after planting, and therefore has a serious problem of synchronicity. In lowland, unimodal rainfall areas, which have dry seasons of over five months, all four of these species also suffer from a serious problem of N volatilization.

Only one of the 19 experiments was done in a way that, according to the principles described in this book, had any chance of increasing long-term soil fertility.[40] In that experiment, leucaena (*Leucaena leucocephala*) and *Senna siamea* were planted in an alley cropping system that received no synthetic fertilizer, but naturally had enough N and leguminous biomass production to have raised yields. Yet yields had dropped.

However, I noticed that the entire decrease in yields had occurred in the first six years. After that, yields had maintained themselves for 14 years. Furthermore, a very heavy mortality rate of both leguminous species had occurred. So with much less biomass than the two tree species should have produced, they had nonetheless maintained yields of about 1.5 t/ha for over a decade. That was about the best one could have expected. But these problems did not explain the drop in yields the first six years. And why had maize yields started at over 2.2 t/ha?

I checked the original source. As I had suspected, the experiment had been carried out on forest land that had never been previously farmed. The low biomass levels of the leucaena and *Senna* could not maintain that original level of fertility provided by the forest, but for 16 years, it did prevent the fertility from dropping to around 0.5 t/ha of maize production, which would be normal in most of the tropics.

But what took me totally by surprise was that this same article also described a second long-term trial, carried out by the same scientists. This second experiment, carried out on land that had been previously farmed and depleted, had experienced *increasing harvests for nearly two solid decades*, before leveling out at 1.5 t/ha. This experiment, however, was never described, or even mentioned, in Bationo's book,[41] even though it appeared in the same article as the experiment he did report. Why Bationo failed to report the second experiment we cannot know, but it very clearly contradicted one of the "key lessons" of his entire book.

In the end, the totality of the information available, including both what Bationo reported and what he failed to report, is completely consistent with the thesis of *Restoring the Soil*: when enough biomass and N is applied to the soil at the right times and in the right ways, it will increase maize yields over the long haul, to a new equilibrium of somewhere between 1.5 t/ha and 4 t/ha.

Finally, and perhaps most conclusively, a very important article appeared in May of 2016 in the magazine *Scientific American*.[42] In this article, called "A Cure for Africa's Soil," some of the world's top scientists from ICRAF in the CGIAR system took credit for work in which, among other things, a farmer's crop production had increased over time by a factor of ten because of the use of leguminous trees. CGIAR scientists were thereby publicly admitting, at last, that gm/ccs are capable of dramatically increasing crop yields over time, and that gm/ccs therefore represent a potential "cure" for soils all over the continent.

Case closed.

The Basic Principles for Productivity and Sustainability in the Tropics

A group of people I worked with in a Honduran NGO called COSECHA, decided in the early 1990s that we needed to find out what kind of agricultural systems would be most productive and sustainable in the world's tropics. The recommendations in textbooks from the temperate world had been leading us in the wrong directions. These books told us that intercropping did not work, but it was clear from fields all around us that it offered significant advantages. They described how to make hay and silage for our animals to feed them when the ground was covered by snow, but we had no snow. Furthermore, making hay was no use at all when we could grow nutritious green forages the year round. They recommended machinery our farmers could never afford, to do jobs, like plowing, that damaged our tropical soils. And then there were the recommendations in favor of using large quantities of fertilizer. It went on and on.

Probably the most misleading idea of all was the conventional attitude toward the long-term productivity of Africa's soils: "Despite the image of luxuriant growth in tropical forests, most of the soils that support that growth are remarkably unproductive."[43] A second author provides the details: "A tropical forest has such a continuing high rate of organic litter decomposition that no mineral pool has time to accumulate. Directly beneath the most recently fallen debris is a heavy, clay-containing mineral soil....As soon as a leaf falls it is decomposed and its minerals are absorbed by plant roots and channeled into the growth of another leaf....When the tropical forest is cut, minerals....leach out of the system and fertility drops sharply."[44]

But the luxuriousness of tropical forest foliage is not some doubtful "image"; it is a fact. If a tropical forest can produce tremendous quantities of biomass on these "impoverished" soils for millennia, why were the same soils unable to produce decent maize crops? How could they be truly "unproductive" when they had been producing, for centuries, some of the highest quantities of biomass/ha on the face of the earth? Maybe the tropical forest's litter layer should not just be dismissed as "fallen debris" sitting atop what is really important. And why was it not possible for a process that quickly turns rotting leaves into new leaves on other forest plants to be utilized by farmers to quickly turn rotting leaves into new leaves on their crops, rather than constituting some inevitable one-way thoroughfare to a sharp drop in soil fertility?

Above all, if smallholder farmers using fallows farmed these soils for over 2,000 years, taking off nutrients in their harvests, how can the experts say that keeping these soils productive will inevitably require large quantities of synthetic fertilizers?

We needed to find some answers. What kinds of farming systems could maximize productivity *and* net profits, as well as maintain a wholesome environment, on these mysterious, seemingly self-contradictory tropical soils? We began traveling all over Latin America in an attempt to answer these questions. We made several stops at EPAGRI in Brazil, of course, but we also checked out scientists in Costa Rica, NGOs in Mexico and Honduras, and innovative smallholder farmers in half a dozen nations. Gradually, over a dozen years, we put together five principles for how to achieve, sustainably and cheaply, the best possible long-term productivity in a tropical environment:

1. Maximize the production of biomass

Conrado Zavala's simple experiment in 1982 (photo 5) impressed on us that SOM could absolutely transform depleted soils. But it also indicated the need for immense quantities of biomass—amounts similar to the annual leaf fall in a 5-year forest fallow. We have found since then that Conrado's results could be replicated all over the world.

But biomass also had to be produced to feed animals, provide firewood and sell in the marketplace. For farmers to prosper, they would have to produce two or three times the biomass they were producing at the time.

Gm/ccs could achieve that. For instance, a system including intercropped pigeon peas and maize, with a 60-day cowpea grown under the other crops and gliricidia trees providing a dispersed shade, can produce three or four times more biomass than maize alone. Competition for nutrients and moisture is largely avoided for several reasons: the four plants feed from different levels in the soil; the 60-day cowpeas will be harvested before the pigeon pea and maize will do most of their growing; and the heavily pruned gliricidia will mostly grow during the dry season. Furthermore, after a few years, the added SOM from the three gm/ccs will have greatly increased both the soil's natural fertility and its moisture content, increasing each of these plants' biomass production a good deal more.

2. Keep the soil covered

But what were we to do with all that biomass? Plowing it all under the soil requires a lot of work, even with animal traction. For farmers using hoes, it is prohibitive. Gradually, we learned of the many advantages of leaving crop residues on the soil surface. Covering the soil protects it from wind, water erosion and excessive heat, the latter of which would volatilize soil N and burn off SOM. Cover also helps maintain soil moisture and provides a constant supply of nutrients, benefiting both crops and soil macro- and microorganisms. We learned that we should maintain soil cover all year round, whether by growing crops (i.e. a green mulch) or by leaving crop residues on the soil surface. The cover is more complete when we intercrop several species at once, including at least one legume and one tree.

3. Use minimum or zero tillage

Valdemar de Freitas in southern Brazil taught us that plowing the soil destroys the soil's structure and causes it to become compacted, leading to the formation of hardpans and claypans and decreased rainwater infiltration. Tillage kills millions of beneficial macro- and microorganisms and buries our soil cover. By decreasing rainwater infiltration, it increases erosion. And in addition to all that, one use of a moldboard plow can destroy as much as 2.2 t/ha of SOM, as opposed to less than 0.9 t/ha through direct seeding.[45] The more we avoid moving the soil, the more productive it will be.

Highly productive zero tillage, however, requires a lot of SOM. SOM keeps the soil loose and builds soil structure, so that plant roots and water can easily penetrate the soil even when it has not been plowed in years. That is why virtually all those two million Brazilian farmers plant gm/ccs in their zero-till fields.

4. Maintain maximum biological diversity

The Brazilians not only regularly use about 60 different species of gm/cc, they recommend that farmers rotate their gm/ccs to prevent build-ups of pests and diseases. Because of the heat in the tropics, pests and diseases proliferate. To achieve good yields over the long term, farmers should rotate and intercrop as many species of food crops and gm/ccs as they can without reducing profitability. We should also maintain as much biodiversity within any gm/cc species as we can, since plant diseases often affect only one or two varieties within a species. Thus, a farmer with biodiversity may be able to continue to use the unaffected varieties of a favored species.

5. Feed the crops through the mulch

Michael Hand, a British scientist in Costa Rica, had been trying for years to improve the yield of common beans in the country's traditional slash-mulch system. The problem resulted from a scarcity of available P, which in turn was caused by a soil pH of between 4.0 and 4.5. But even when he applied ample quantities of P fertilizer, yields stubbornly refused to budge above 0.5 t/ha. Eventually, Michael tried just throwing some P fertilizer on top of the mulch. Yields immediately jumped to between 1.5 and 2.5 t/ha.[46] Less work brought more productivity!

Tropical soils often have serious nutrient deficiencies and imbalances. Some soils are extremely acidic, with virtually no available P and toxic levels of aluminum. Still others have impenetrable hardpans and claypans. To be honest, these soils have become hostile environments for feeder roots. On the other hand, mulches on top of those soils usually contain adequate levels of nutrients in well-balanced quantities that are highly accessible. They are a much better source of nutrients than the soils beneath them. And feeder roots are perfectly willing to grow upward if that is where dinner is being served.

If we take seriously the nutrient access idea, then, a mulch can feed plants very well, as long as it remains moist and biodiverse.

One day in a personnel meeting, after a dozen years of developing and testing these five principles, COSECHA's personnel were discussing the synergy between them. For instance, it was clear that zero tillage and lots of biomass would lead to soil cover, and that, in turn, would lead to plants' feeding from the mulch.

Suddenly the most important relationship between these principles became stunningly clear to us. These five simple principles merely describe the way tropical forests have maintained their extraordinary biomass productivity for millennia. That is, all we had discovered during our 12-year-long odyssey of discovery was something we should have guessed all along: for smallholder agriculture to be both highly productive and sustainable for generations, even on supposedly nutrient-poor soils, it must imitate the highly productive, millions-of-years-old tropical forests.

After all, no one ever needs to plow a tropical forest. Tropical forests are world-famous for their tremendous biodiversity and extraordinary biomass production. They keep the soil covered all the time and, perhaps strangest of all, the forest's trees feed almost entirely from the forest's litter layer (its natural mulch). That is, the trees' leaves do not drop onto bare soil, but onto the litter layer. The trees' roots form a dense mat immediately below and inside that litter layer. This mass of roots catches the fallen leaves' nutrients before they ever touch the soil and immediately after they have decomposed enough, transports them back up into the bushes and trees. Never touching the soil, the forests' nutrients never become tied up and unavailable. Recycled within weeks, these nutrients produce much more tree biomass than if they spent a dozen years in the soil before being recycled.

This, then, is how tropical forests on seriously nutrient-deficient soils manage to produce tremendous amounts of biomass. If smallholder farmers are to produce good harvests year after year at minimal cost, they would do well to imitate the forests. Millions of farmers are now doing exactly that.

The tropical world is now developing and refining agricultural systems specifically designed for tropical environments. By imitating the tropical forest, it has created systems that can remain highly productive for centuries. In so doing, smallholder farmers can very cheaply produce up to 4 t/ha or more of maize for decades.

Now *that* is good news!

Interestingly enough, during the last fifteen years, a growing number of farmers in temperate countries, such as Gabe Brown in North Dakota, USA, and Christine Jones in Australia, are achieving better than average productivity at greatly reduced costs (and therefore much higher than average net incomes) by using some of these same principles. Most notably, they are using minimum or zero tillage, maximizing biomass production and maximizing plant biodiversity, which also results in increased soil cover. In the end, it just may turn out that most of these previously strange principles are appropriate for far more than just the tropics.[47]

Also of great interest is that these farmers from temperate climates, as well as a few pioneering scientists like David Johnson, are finding that the integration of animals into such systems can improve yields even further because of their impact on beneficial micro-organisms in the soil. This practice would also, of course, imitate another characteristic of natural forests. Someday, once we have had time to confirm more widely the benefits of integrating animals, we may have six basic principles for maximum productivity and sustainability—everywhere!

The 3-Tiered System

Even so, our learning curve continues its steep climb. Increasingly, we are realizing that an ideal gm/cc system for almost anywhere in the lowland tropics is one that consists of three tiers of biomass. One example would be our system of growing maize with intercropped pigeon peas, an undergrowth of crawling cowpeas and gliricidia-based dispersed shade. With such a system, a farmer would produce more than enough maize, cowpeas and pigeon peas to feed the family and pay for all the production costs; make a good income from whatever the family did not eat; produce a good quantity of firewood; and provide enough biomass to triple the farm's maize productivity within five years. The system would also keep the soil covered, never require any tillage, produce a maximum of biodiversity and reduce wind speeds. The more than 3 t/ha/year of carbon sequestered as humus, as well as the long-term fertilization of the soil and the almost complete resilience to drought, all achieved within about five years' time, would constitute additional *free* by-products of the system.

A 3-tiered system would include:

1) dispersed or intermittent shade, both of which are provided by trees that are spread out across a field;

2) crops, as well as gm/ccs, like the pigeon pea, that grow alongside the main crops; and

3) crawling crops or legumes that stay on or near the ground, providing soil cover that protects the system from N volatilization and the evaporation of soil moisture. Many climbing gm/ccs, such as lablab beans, would, by themselves, provide both the second and third tiers.

Although today most of the dispersed trees in our trials are still below their full, mature size, such 3-tiered systems are producing very positive results. These 3-tiered systems do better than any others in supplying benefits in greater quantities than do simpler systems. Once we have a little more evidence, we may well add another basic rule to the above list: include dispersed or intermittent shade in the system.

No one need remind us that adding trees to our fields represents a very obvious, giant step closer to imitating a natural forest.

Gm/ccs and Conservation Agriculture

Conservation Agriculture (CA) is now widespread in much of Africa and South America. The relationship between CA and the five principles of maximizing productivity and sustainability in the tropics is very simple. Our principles #2 (keep the soil covered), #3 (use minimum or zero tillage) and #4 (maintain maximum biodiversity) are identical to the three principles of CA. Therefore, to maximize CA's yields, we only need to add to CA the principles of maximizing biomass production and feeding the crops through the mulch.

Gm/ccs alone can both dramatically increase biomass production and provide the biodiverse mulch from which crops can and will feed. That is, adding gm/ccs to CA can convert CA into a system that maintains or increases soil fertility and moisture retention for centuries. Furthermore, it will do so at much less cost than any system that includes carrying mulch material onto the fields or using compost or fertilizer, as CA proponents recommend in much of Africa. And gm/ccs can eliminate the three major constraints that limit how much land African farmers can dedicate to CA: the excessive labor involved in carrying mulch material to their fields, the limited amounts of available animal manure, and the costs of using mostly fertilizer, compost or animal manure to raise their yields.

The incorporation of gm/ccs into CA can make a huge difference in the acceptance and productivity of CA. This factor is illustrated on a continental scale by the differences in the acceptance of CA in Africa and Brazil.[48] In Africa, very few farmers use gm/ccs with their CA. And CA, though widespread, has seen very little spontaneous spread from farmer to farmer.[49] Also, very few African farmers ever do CA on more than 0.3 ha, and yields of CA plots seldom pass 2 t/ha unless other sources of OM or synthetic fertilizers are used. Even then, yields seldom reach 5 t/ha.

In Brazil, by contrast, farmers consider gm/ccs a vital part of CA. Most farmers would not think of practicing CA without gm/ccs. In this case, over 2 million farmers now practice CA on over 25 million hectares, for an average of about 12 ha/farmer. And yields often reach 6 to 8 t/ha (with the use of small amounts of fertilizer). At these levels of productivity, it is hardly surprising that the spontaneous spread of CA in Brazil has been impressive, indeed.

How Much Organic Matter is Required to Increase Soil Fertility?

How much OM must a farmer apply to the soil to maintain, or even improve, the soil's fertility? Surprisingly, I can find no evidence that this question has ever been scientifically researched, even though it is absolutely essential to understanding how to manage low-cost, sustainable agriculture.

Even if researchers did try to establish certain standards, no exact figure could be established. The exact amount would depend on a series of factors that vary from one place and one gm/cc system to another: the chemical characteristics of the OM being used; the density of nutrients in that OM; the quality of the soil; the factors in the environment that limit productivity; the amount of erosion; the climate; the ways the biomass and soil are being managed; etc.

Nevertheless, there is a desperate need for at least an educated estimate. Such a figure would allow us to have a rough idea of whether or not any particular gm/cc system would be able to maintain or increase a field's fertility, productivity and OM content.

Of the more than 100 gm/cc systems I have personally observed, those that can maintain soil fertility steady over many years are those that supply somewhere between 20 and 25 t/ha/year of leguminous OM (green weight), in addition to normal maize crop residues. Larger amounts of OM would, of course, increase soil fertility.

This figure is admittedly a rough estimate. It refers only to average above-ground leguminous biomass. I have seen mucuna, for instance, that produced as little as 20 t/ha of biomass, and as much as 105 t/ha. Such figures will vary because of the maturity of the gm/cc, number of seeds planted/ha, broadcasting of seed vs. planting in holes, the opportunity to climb rather than just grow along the ground, the amount of shade cast on the legume, the fertility of the soil, weed competition, climate, etc. Nevertheless, I have never observed a case in which a gm/cc system flagrantly violated this rule of thumb by, for instance, producing only 15 t/ha of biomass and still maintaining the soil's fertility.

The following chart summarizes the analysis behind this estimate (weights are green weight). The numbers refer to the amount of biomass still remaining in the field when the following crop is planted.

Name and number of system	Approximate average amount of OM produced/year	Does it usually maintain or improve the soil fertility?*
Systems that increase soil fertility:		
3rd maize/mucuna system (S52)	60 t/ha	Fertility improves
Maize/sweet clover (S103)	40+ t/ha	Roughly maintains, but also feeds grazing animals
1st maize/runner bean system (S97)	50 t/ha	Improves
1st maize/lablab bean system (S15)	40 t/ha	Improves
Systems that more or less maintain soil fertility:		
Maize/fava beans (S102)	25 t/ha (depending on the density of the legume)	Maintains fertility
Millet/native dispersed shade/ short-cycle cowpeas (S69)	20 t/ha	Roughly maintains
Tephrosia fallow (S8)	60-80 t/ha, once every 4 years	Maintains
Systems that fail to maintain soil fertility:		
2nd maize/tithonia (S45)	15 t/ha (depends on spacing of hedgerows)	Fertility decreases
2nd hedgerow/tephrosia (S47)	10 t/ha	Decreases
2nd potato/tarwi (S107)	5-10 t/ha (depends on the size of the field)	Decreases
Rotation/common peas (S115)	5-15 t/ha	Decreases

*Systems that supply 25 t/ha/year or more of OM to the soil will gradually increase soil fertility until harvests reach somewhere between 3 and 5 t/ha/year of maize, and then yields will level out. All figures given are for systems that use no synthetic fertilizer.

To most people, 20 to 25 t/ha sounds like an extraordinary amount of OM. It is. Those who have promoted compost will immediately realize that virtually no smallholder farmer uses even half this quantity of compost, except occasionally in small-scale vegetable gardens. Thus, compost alone will almost never allow a farmer to maintain his or her soil fertility by itself. But a good number of gm/cc species can produce from 3 to 6 kg/sq m of biomass, even when intercropped. Since a hectare contains 10,000 sq m, these gm/ccs are producing from 30 to 60 t/ha/year of biomass.

The Role of Soil Biology

The Nutrient Quantity Theory has heavily influenced most people's concept of how the soil works. For instance, what we call a "soil analysis" almost never tells us anything about the levels of biological activity in the soil, what the soil structure is, whether the soil has a hardpan and how deep it is, the soil's water-holding capacity, or even the amount of OM in the soil. Only rarely does a soil analysis even tell us the total amount of nutrients in the soil or the percentage of nutrients that is available to our crops. It tells us how many nutrients the soil has that are chemically accessible to the plant, and how much fertilizer we presumably need to apply. But we all know that keeping a soil productive requires a good deal more than merely maintaining a certain level of accessible nutrients.

Furthermore, despite our having known since at least the 1960s that every tablespoon of fertile soil contains millions of micro-organisms, and that higher quantities of micro-organisms correlate with both higher natural fertility and higher accessibility of nutrients to our crops, very little was done until about the very end of the 20th century to investigate the role of soil micro-organisms in the care and feeding of crops.

Thankfully, today we know a lot more.[50] We know, for instance, that mycorrhizal fungi can multiply several times the ability of roots to absorb nutrients and moisture. We know that soil fungi play an essential role in making nutrients accessible to plants. We also know that synthetic fertilizers kill off mycorrhizae. In addition, increasing evidence indicates that the quantities of synthetic fertilizers previously recommended by soil analyses were far higher than the soil needed to maximize productivity. In fact, the experience of some well-known farmers has provided evidence that even average levels of productivity in the US, and much higher than average US net incomes, can be achieved decades after the use of all synthetic fertilizers has been terminated.[51] We desperately need to know more about these issues.

In the meantime, we know that with more carbon in the soil, the number of microorganisms increases dramatically. We know that as millions of farmers have increased the OM and carbon content of their soils, their soils have become more fertile and more capable of holding moisture. We still do not know *how* all this happens in the mysterious, largely uncharted depths of the soil, but I know of no one who still doubts it is happening after having visited a few farmers who have succeeded in restoring their soils.

Nitrogen Dynamics

The natural N in our soils comes from the air, almost all because of rhizobia that fix atmospheric N. These organisms are particularly important in the tropics because N is the main nutrient that limits how much our crops produce. No matter how many other nutrients are available in our soils, if N is the limiting factor (i.e. the nutrient most lacking in the soil as a proportion of what plants need), the amount of N in the soil will control how much farmers harvest.

N is particularly deficient in smallholder farmers' tropical soils for two reasons: 1) because N moves so easily through the soil. Rainwater is constantly washing it down through the soil, beyond the root zone of our crops, and 2) N that is on top of the soil or just beneath the surface is constantly being burned off (i.e. "volatilized"). This high rate of volatilization means that farmers in the lowland tropics need to add more N to their soils than do farmers in temperate climates.

The movement of N within a crop plant is complex, fascinating, and important to understand. Early in the plant's life, most of the available N from the roots and rhizobia moves to the leaves. As the flowers, pods and seeds start forming, most of the N moves again to these reproductive parts of the plant. In fact, so much N moves that by the time the seeds have begun growing, the root nodules have usually disintegrated. If the plant lacks enough N, the older leaves on the plant will turn yellow and may even die. This movement of N from the roots and leaves means that the leaves that die and fall to the ground contain only a fraction as much N as do green leaves. Pruning green leaves is therefore very important for maximizing the impact of gm/ccs—especially fertilizer trees.

Phosphorus Dynamics

Whereas with N the challenge consists of adding more N than what is being lost, the primary challenge with P in the tropics is to use efficiently what we already have, by keeping it accessible to our crops.

Since 1990, the world's leading soil scientists have been predicting that within 15 to 20 years, Africa would suffer a major soil P crisis, causing widespread malnutrition and hunger.[52] The only problem is that today, 28 years later, visible P deficiencies in crops are still almost non-existent across Africa, except in areas with a soil pH approaching 4.5 or less. Even after 27 years, it is still N, not P, that is limiting food production for the vast majority of Africa's and Latin America's smallholder farmers.[53]

Scientists have given two major reasons for this purported soil crisis.

The problem of soil analyses

Farmers throughout the developed world know they should apply chemical P in bands rather than spreading it evenly across the soil. They do so because a huge percentage of the P that comes into contact with the soil, as opposed to being in contact with other molecules of P, will rapidly become tied up (i.e. chemically unavailable to plants). In tropical soils, the amount of total P that is available to crops (i.e. not tied up), usually ranges from virtually nothing to about 3 % of total P.

However, SOM can dramatically improve the availability of P. First, it buffers pH, so that acid soils become significantly less acid, thereby freeing up a lot of P. More important, as decomposition of the SOM occurs, all of its P will become available to plants at one time or another. Therefore, if the P in our crops or trees falls on a biodiverse mulch with plenty of crop and tree roots growing up into it, plant roots will absorb the P before it can touch the soil and become tied up—thereby utilizing not just perhaps 10% of it per year, but close to 100%.

For centuries, African farmers have had fair amounts of OM in their soils because of fallowing. But as population pressures over the last 30 years have reduced the average smallholder farm size to less than 2 ha, Africa's smallholder farmers have had to reduce their fallow periods from 15 years down to 2 to 4 years or, in most cases, to nothing. Since fallowing produced far larger quantities of SOM than anything else farmers do, this slow death of fallowing has caused a major, unprecedented reduction in SOM levels all across sub-Saharan Africa.

So why have soil analyses indicated a major drop in P over those same 30 years? Simply because soil analyses normally do not measure total P; they measure available P. They do not measure the 97% of the P that is tied up in the soil, but only the 3% that is available to crops. As the SOM in Africa's soils decreases, the availability of soil P also decreases, even if total P does not. As a result, soil analyses have (rightly) indicated that *available* P has decreased. But many people have (wrongly) assumed that *total* P has also decreased. The decrease in available P should not lead us to believe that total P in the soils of Africa's smallholder farmers is decreasing, much less that we are facing a continent-wide soil P crisis.

The problem with nutrient assessment calculations

The second reason for the predictions of an African soil P crisis was the large off-take of P calculated by the "nutrient assessments" done across Africa. These assessments tried to calculate the net impact caused by all the known gains in soil P as opposed to all the losses on smallholder farmers' fields. Very simply, these assessments are seriously wrong because they are built on a major misunderstanding about African smallholder farmers. In the United States or Europe, it can be quite accurate to calculate nutrient depletion by including all harvests as nutrient losses and totally disregarding what is happening to the nutrients in human waste. After all, northern farmers' human waste is flushed down toilets, to be enclosed in septic tanks or piped miles away from their farms.

On the other hand, the vast majority of African smallholder farmers consume most of what they produce. Furthermore, most of them produce only enough food each year for from 5 to 10 months of their families' consumption. They are therefore major net importers of food. Given that smallholder farmers import into the village anywhere from 20 to 60% of the food they consume (plus added food to replace whatever food they sold right after the harvest), the net transport of P *into* the village would be equivalent to somewhere between 10 and 100% of the total P being harvested. (Commercial crops like cotton, sesame or vegetables will affect these calculations, but they are usually fertilized with synthetic fertilizer containing a lot of P.)

Using conservative figures, at least 20% of the family's human waste goes directly into the fields. Most of the rest goes into a family's shallow, hand-dug latrine. The roots of nearby trees and bushes easily capture these nutrients and sooner or later drop the leaves, pods and flowers into or near the family's courtyard. Sweepings from village compounds are carried back out to the fields. Insects and microorganisms also transport these nutrients directly from the latrine into the compound or the fields. Since P is immobile in the soil, it presumably stays at the bottom of the latrine until transported to some other location by plants or animals, from where it almost inevitably winds up back in the fields. Where else could it go?

Of course, if farm families live in a small town far from its fields, this nutrient cycle may be cut, but the vast majority of Africa's smallholder farmers live in small villages relatively near their farm plots.

Unfortunately, the nutrient assessments done in Africa completely fail to mention families' consumption of most of the food they produce, the import of food into the villages and the destination of the resulting human waste.[54] Since harvest losses usually are the largest single source of nutrient loss in these assessments, these omissions are the elephant in the middle of the room that no one sees. These analysts are either 1) assuming that all the food produced on smallholder farms is exported from the farm, 2) assuming that the nutrients in farm families' human waste just disappears, presumably down a flush toilet, or 3) purposely distorting their results. Anyone who has ever spent a few days or nights in African villages knows that people there eat a large majority of the basic grains they produce, and flush toilets there are about as common as two-car garages.

Inasmuch as smallholder farmers are almost all net importers of nutrients into their villages, it is highly likely that they are *increasing* the total P in their soils every year, and they will not have to worry about any so-called P deficit until that distant, happy day they become net exporters of food. By that time, they may well have prospered enough to more easily afford the replacement P they will need.

One more advantage of zero till, mulch-based agriculture: the recycling of P

Furthermore, even if stocks of P in African soils were hovering somewhere near crisis levels, the dynamics of mulch-based agriculture mean that the situation is not as dire as one might think. In temperate countries, only around 10% of the P in a fertilizer is used by the crops the first year after application, and a good deal less in succeeding years.[55] Thus the time that P spends in the soil before it is used by a crop averages somewhere over 15 years. In the tropics, that time frame may be stretched even longer because of the soil's extreme pH's and low productivity.

On the other hand, if leaves drop from a tree or bush onto a litter layer in a humid tropical forest, they will take perhaps 2 to 6 months to mineralize and the P will almost immediately be absorbed by roots and transported up into another tree or bush. Most of it will go to the leaves, pods, flowers and seeds of these plants, and then drop to the litter layer within an average of another 6 months or so.

A cropped field will mimic the tropical forest in this respect if the farmer maintains a thick, biodiverse mulch. Farmers using adequate gm/ccs and zero tillage will normally be doing precisely that. Their crops' feeder roots will then grow up into the mulch, and the P in the mulch never need touch the soil. If the system is based on annual crops, an average of around 90% of the P will recycle through the mulch and back into another plant within one year, instead of every 15 years. The result, obviously, is that the P in a permanently mulched system will feed crops about 15 times more efficiently than in a conventional field!

In conclusion, not only is there a good deal more P in Africa's smallholder soils than nutrient assessments or soil analyses would have us believe, but farmers can produce about 15 times as much biomass with a given amount of P than we previously thought. It may be true that you cannot get something from nothing. But farmers *can* get an awful lot more than we previously thought, from very little, once they have learned how.

The on-the-ground evidence

Aside from theoretical considerations, what on-the-ground evidence do we have that the P crisis is not occurring? For one thing, the predicted disaster has not arrived, nor is it even close. But even fairly well-thought-out long-term predictions may err by a few years. What other evidence do we have?

Probably the most closely studied cases of nutrient dynamics in a gm/cc system are those of the 3rd maize/mucuna system (S52) in Honduras and several systems in Brazil. In the case of S52, over 5,000 farmers in Honduras alone have been growing maize and mucuna on the same fields every year for at least 30 years. Most of these farmers have never applied synthetic P to the soil. Yet, according to a Cornell University doctoral dissertation authored by Bernard Triomphe, after 20 years of harvesting maize at rates of 2.5 to 3.5 t/ha, Triomphe found no yield response when he added P fertilizer to these soils.[56] There was no P crisis in the making. In Brazil, no P shortage has been found, either.

More generally, during a study I carried out in late 2009, I traveled through significant areas of six African nations, interviewing hundreds of farmers. In the process, I observed all the agricultural practices and conditions as I traveled along the highways and dirt roads, in order to triangulate that information with what I heard from farmers. I saw tremendous evidence that yields were diminishing, primarily because of nutrient deficiency. But N was clearly most deficient in at least 97% of the tens of thousands of fields I observed. Wherever one goes (except perhaps the rainforests of central Africa), one can observe the yellowish, stunted maize and millet fields that indicate an N deficiency. On the other hand, the purple-edged leaves associated with a P deficiency showed up in less than 3% of the fields. P may still, of course, be lacking to some extent, but any P deficiencies there might be are being masked by much more serious N deficiencies.

In my travels across more than 60 developing nations, I have observed the same thing: fields everywhere without gm/ccs commonly show N deficiencies, but only rarely deficiencies of P. Yields are trending downward if gm/ccs are not being used, but the main problem is not P. Gm/cc systems often prove this point most dramatically. Outside of Brazil, most farmers who use gm/ccs use no fertilizer, and they harvest double or triple what they would without gm/ccs. Yet except where the soil pH is below 4.5, I have never observed a serious P deficiency with any of the systems described in this book. This is true even where farmers have applied no P for decades. The maize/mucuna phenomenon in Honduras is not an exception; it is the rule.

In conclusion, net off-take of P is not occurring widely on African smallholder farms. Even without P fertilizer, no P crisis exists, nor will it exist, until sometime well after African smallholder farmers become net exporters of P. Even then, by implementing the above rules for managing tropical soils, they can postpone the expensive day when they will need to buy replacement P.

Potassium Dynamics in the Soil

What we need to know about the flow of potassium in soils is quite simple. First, potassium is much more stable in the soil than is N, except when the soil is quite sandy. In that case, potassium can easily be washed down deep into the soil, creating a deficiency of potassium in crops' root zone. Second, crops that contain a lot of sugar (such as sugarcane) or starches (such as white potatoes) require a lot more potassium than do most crops. These crops can therefore cause potassium deficiencies in the soil. In either of these cases, animal manure or small amounts of potassium-rich synthetic fertilizer are probably the best ways to solve the problem.

Competition for Biomass with Grazing Animals

Some agronomists have maintained for years that gm/ccs cannot work in semi-arid areas, and even if they did, the use of vegetation to increase soil fertility would compete directly with the feeding of cattle. The idea that gm/ccs cannot work in semi-arid areas has now been dramatically disproven by the success of FMNR and other gm/cc systems in a dozen nations.

Still, almost everywhere, free-grazing animals present one of the most difficult problems to deal with when selecting a good gm/cc system. Farmers definitely want to use gm/ccs that are edible, but plants that are edible for humans are also edible for free-grazing animals. Furthermore, gm/ccs are usually the most nutritive of all available biomass, so grazing animals will often attack them first. The technically simplest, and ecologically best, way to overcome the problem is to produce more biomass than the animals can eat. This solution is often used, but in drought-prone areas, producing that much biomass can be a challenge.

The solution programs often try first is to convince the villagers to prohibit free-grazing. This is often difficult, but it is the approach that has been used to achieve most of the adoption of FMNR on over 20 million ha across Africa. Where this has worked, it has opened up all sorts of additional possibilities for better productivity and environmentally sound land management. Lablab beans can make such a decision profitable for cattle owners and land-owners, because of their high nutritional value, their high palatability, and their ability to grow through the dry season in most climates. Even many large-scale commercial cattlemen in Honduras plant maize and lablab together in order to graze their cattle on lablab during the dry season, rather than let them free-graze (see S17).

However, convincing people to control their animals is not easy. It is especially difficult if the number of animals is heavily skewed toward the better-off members of the community, because the richer people obviously gain the most from free-grazing, and they tend also to be among the community's more influential decision-makers. Also, even if one village decides to control its animals, free-grazing cattle from a nearby village may finish off the gm/ccs. This is a particular problem with lablab beans, because they tend to be the most palatable of all the fodder species around.

One of the most important factors to remember where free-grazing cattle exist is that they, like most animals, are priority grazers. They will start by eating one group of species, and will not touch anything else until they have largely eliminated the preferred species. As they get progressively hungrier, they will eat a wider and wider selection of species.

For instance, in most areas of the developing world, cattle will not touch jack beans. In the Sahel this was also true, until they became really hungry toward the end of the dry season. At this point, they suddenly finished off all the jack beans in the area. Some books claim that because tephrosia is poisonous, animals will not eat it, but when we planted tephrosia in central Mali, the cattle eliminated it, too, just before the rains started. Whether the animals will eat a specific gm/cc is often something one must try in each locality, remembering that during droughty years, animals will consume a wider range of species than during better years.

A much easier, very inexpensive solution in most areas where farmers have long-term usufruct rights to the land (as is true in most of the Sahel) is to plant live fences around fields with gm/ccs. Most species used in Africa are either very slow-growing, unable to withstand pressure from hungry cattle or not very long-lasting. The recently discovered monkey thorn tree (*Acacia galpinii*) may be a very good solution to these problems. It is very strong, forms a good fence within one year, and lasts at least forty years, with just one annual pruning to keep it from growing too tall. Furthermore, it is a good fertilizer tree—its prunings increase the leguminous biomass available for keeping the soil fertile.

If nothing else works, fodder trees may be the only solution. Gliricidia will grow out of reach of animals, and therefore can be a wonderful source of dry-season fodder that is restricted to the owner's animals. Even nomadic herders like the Fulani will not normally cut down branches of someone else's gliricidia, because of cultural prohibitions in desert areas against cutting other people's trees.

Intercrops vs Rotations

One major choice to be made in using gm/ccs is whether to intercrop them or rotate them with farmers' other crops. Traditionally, in temperate countries, rotations are heavily preferred. This is in part because intercrops are very difficult to manage with machinery.

The preference for rotations also results from some unique characteristics of the world's temperate zones. It is no secret, for instance, that the amount and intensity of sunlight is much greater in the tropics than in temperate zones. This more intense sunlight obviously means that plants growing in the partial shade of adjacent plants will still receive a fair amount of sunlight in the tropics. Less widely recognized, the sunlight in the tropics comes from more directly overhead. This means that whereas in the temperate zones, the sunlight is coming in at an angle that causes the shorter plants in an intercrop to live in almost total shade, the shorter plants in the tropics will often receive a fair amount of sunlight.

Lastly, in the lowland tropics, the midday heat is so intense that most unshaded plants simply stop growing for a couple of hours, thereby reducing their productivity. A light shade, whether from a dispersed shade or from nearby intercropped plants, may actually increase the productivity of the intercropped species. Thus, because of increased sunlight, the tropics enjoy a tremendous advantage over temperate areas in terms of increasing productivity through intercropping. In fact, widespread experimentation in the tropics has shown that, in general, intercropping can increase productivity of a given plot of land by up to 70% over monocropping (when, of course, the productivity of both crops is added together). One of the reasons for this increased total productivity is that legumes have different root structures from grassy subsistence crops, so they do not compete for nutrients from the same layers in the soil.

Several other important considerations favor intercropping beyond those taken into account in the above experiments. Many viny gm/ccs, such as lablab and mucuna, produce up to 50% more biomass just because they have something to climb. Intercropping, especially with legumes, helps control many insect pests (e.g. the fall army worm), plant diseases, parasites (e,g, striga and nematodes), and weeds in general. Furthermore, the general rule of maximizing biomass production per hectare provides a long-term benefit of intercropping not taken into account in intercropping experiments.

As we learn more about gm/ccs and their use, we will also find ways to intercrop more than just one legume in a grain field, increasing the benefits of intercropping even more. The most obvious case is that of dispersed shade and intermittent shade systems. These trees can be combined with virtually every other low-altitude gm/cc system we presently use, including pasture production, adding one or more new species to the system. The 60-day cowpea, also little known or used, provides the possibility of adding an additional intercropped species, because it matures and ceases to occupy space before the other crops grow large enough to occupy that space. Crawling cowpeas are opening another whole range of possibilities for adding an additional intercropped species.

In terms of limitations to crop density in intercropping, experience shows that competition for nutrients is normally not nearly as problematic as competition for soil moisture. During droughty years, crops planted at high densities can suffer a 15 to 30% decrease in productivity because of crop density in areas where no decrease is experienced during years of normal rainfall. The importance of nutrient competition will become even less important once we have built up our mulch layers. As the mulch layer grows and soil moisture retention improves, both factors will allow farmers to increase their plant populations and biodiversity even further.

Nevertheless, there are limits. Even in the tropics, competition for sunlight is often the limiting factor for plant density in intercropping systems.

Native Species or Exotics

Another vexing issue we must often face when working with gm/ccs is whether to use a native species or an exotic species, often from another continent. Native species are usually more resistant to local pests and diseases, ecologically more desirable, and better known by local farmers, but exotic species may have characteristics that are desirable, or even crucial, in a given situation.

In fact, a tremendous number of exotic species have been used in agriculture for centuries, largely without mishap. If, for instance, we tried to eliminate exotics from United States agriculture, we would have almost no agriculture left; not a single even moderately important crop species in the United States is native to the country. The situation is similar elsewhere. No longer using exotics would deal an incredible blow to the agricultural productivity of virtually every country in the world.

Even so, whenever we agronomists consider introducing an exotic species, we should be very careful. This is especially true for species that grow vigorously with very little assistance. At a minimum, when we do introduce a new species that is not already used in multiple countries, we should know for certain how that species can easily be eliminated from an area, without the use of herbicides. Nearly all the species recommended in this book are known to be quite safe because they are already being grown in many countries without any problem. Those that have caused problems anywhere are listed as species that should never be introduced into new areas.

How Much Carbon Can Gm/ccs Sequester?

The potential for gm/ccs to sequester carbon in the soil are tremendous, and are greatly underappreciated.

The following calculation is based on a moderately good gm/cc system—a maize crop with intercropped pigeon peas and an understory of Lojy Be cowpeas. This is not an ideal system, since it could be improved very simply by adding a dispersed shade of gliricidia. It is also based on current knowledge of gm/ccs, rather than future potential. Despite these limitations, this calculation can give us a general idea of what is possible.[57]

Part of plant After 10th year:	Biomass (t/ha dry weight)	Long-term organic carbon
Pigeon pea above-ground	7.5	2.2
Pigeon pea below-ground	3.2	1.4
Lojy Be above-ground	2.5	0.7
Lojy Be below-ground	1.0	0.4
Maize above-ground (half of biomass)	1.8	0.5
Maize below-ground	1.5	0.6
Total:		**5.8**

That is, 5.8 t/ha X 65 years between 2035 (when this system would be well-established) and 2100, equals 377 t/ha. If this were done on all the 225 million ha that smallholders are calculated to have (about 12 % of the world's agricultural land), the total C sequestered would be close to 85 Gt. This represents just over 10.5% of all the 810 Gt of carbon the entire world will need to sequester by 2100 to reach the goals of the Paris Accords. If the world's large-scale farmers and ranchers could sequester similar levels of carbon/ha, all the world's farmers and ranchers would be sequestering over 85% of the carbon needed to reach these goals. Ranchers grazing cattle on pastures should be able to double that level of carbon sequestration.

These figures are, of course, approximations. We do not know very accurately, for any given farming system, how much of the carbon placed on or in a soil will become long-term carbon. Other estimates tend to be lower than mine, but they are based on temperate climate agriculture, where biomass grows much more slowly, and were done by scientists who are unaware of the amount of biomass produced by the gm/cc systems described in this book. I am quite confident that these figures are, if anything, rather conservative, because I have triangulated them with figures from 40 years' growth of the 3rd maize/mucuna system (S52) and similar cases I have worked with closely in Honduras. The actual build-up of topsoil from these cases produces figures higher than those given in the table.

What will it cost humanity to sequester all this carbon? A whole lot less than storing carbon in caves or subsidizing wind energy or spraying our atmosphere with anti-global-warming particles. In fact, it will cost nothing to the farmers themselves, because sequestering that carbon is a free by-product of their increasing their own agricultural productivity and net incomes.

Chapter 5. Building Resilience to Droughts

An Unexpected Benefit

In November of 2011, during one of the worst droughts that Mali has suffered in recent memory, I drove with several colleagues from the nation's capital, Bamako, northeast into the affected area of the Sahel. It was harvest time, except that for hundreds of thousands of farmers all around us, there was not going to be any harvest. For 500 km, we traveled through fields of badly stunted, knee-high maize, sorghum, and millet that were not going to produce more than a few kilos of grain/ha.

About 10 km northeast of the Malian town of Koro, near the Burkina border, we started seeing a strange sight. In every field, there were somewhere between 80 and 150 trees/ha—native trees of many species. Strangest of all, every one of them was shaped in a funny inverted cone, like huge funnels from a kitchen. Furthermore, the people were harvesting and hauling back to their villages a bumper crop of millet and half a dozen other crops. Even with over 40 years' experience in development work, I had never seen anything like it. I had no idea what was going on. I saw no evidence of irrigation or water-harvesting of any kind. Yet these people from some 20 or so villages of the Dogon ethnic group, using technology they had developed totally by themselves, were taking in the only decent harvest for 150 km in every direction. Somehow, they had created a veritable oasis in the middle of a desert.

What had the Dogon people done to bring about such a dramatic development?

After we had said our elaborate, obligatory greetings, asking how everything was going right down to the people's cattle, I asked one of the Dogon leaders about those funny, funnel-shaped trees. Good teacher that he was, he started by asking me a question to illustrate his point: “What happens to crops that are planted right under a mango tree, with its dense foliage down close to the ground?”

“The crops die.”

“Why?”

“Because they never get any sunlight.”

“Exactly. But if the leaves of a tree are high off the ground, their shadow gradually moves across the field as the sun travels across the sky. That way, all the crops get good sunlight at least part of the day. And that way they all grow very well.”

The Dogon people were using what we now call “intermittent shade.” At the same time that it was protecting the crops from the excessive heat, this intermittent shade was reducing both the evaporation of moisture from the farmers’ soil and transpiration of moisture from their crops. All that moisture is thereby no longer being sucked out of the farmers’ soil. This means that during a drought, crops protected by intermittent shade will have more soil moisture and will produce a lot better. In addition, the leaf fall can increase crop growth another 20 to 50% because of the mulching effect, the higher rainwater infiltration, the added SOM and, if the trees are leguminous, the added N.

I had never seen a case of intermittent shade before that November day, because as far as I know after all my wanderings, no agronomists or other farmers in the world had ever thought of it. The Dogon people invented it. This little known, often looked down on “tribe” from the middle of nowhere (in actuality, not that far away from Timbuktu), has given the world an incredibly valuable technology that we are now hoping to spread across the tropics. Intermittent shade raises yields, but it can also provide firewood and fodder, so forests do not have to be cut down. And it will help defend farmers on three continents, as well as all the rest of us, from global warming, because carbon will not only be increasingly sequestered in the soil, but also in the trees themselves.

We still need to do side-by-side tests to compare intermittent shade with dispersed shade, but we suspect that, of the two systems, intermittent shade will be slightly more advantageous over-all. In the case of dispersed shade, the pruning of the trees is somewhat easier, but in the case of intermittent shade, more trees can be grown in a hectare and each tree can have more biomass, thereby providing all the advantages of extra OM, fodder, etc. It could be that intermittent shade will be preferable on farms where more labor/ha is available or people have more animals/ha, whereas dispersed shade would be better where animals are rare or soil fertility is good enough that the extra biomass is not that crucial.

The only additional cost in an intermittent shade system is the labor of pruning the trees in that funny shape once a year. Except that trimming those trees saves African women the huge job of walking 4 km to a piece of forest and climbing the trees to cut firewood and then haul it all the way back to the village. All in all, these trees in the fields save labor. And they cost nothing to grow. After all, they were already there before the Dogon people ever arrived.

The Dogons were doing more than just growing shade trees in their fields. They were also planting a series of drought-resistant gm/ccs they had pulled together themselves, from bambara groundnuts (*Vigna subterranea*), peanuts and 60-day cowpeas to *fonio* (*Digitaria exilis*), an edible grass that produces a tremendous mass of roots. All of these crops had been increasing the OM content of their soils for decades, and because of the decreased volatilization under the tree shade, they had also been increasing the amount of N in the soil.

But How Can This Be?

How is it that a lot of OM and a little shade can enable crops to grow right through a serious drought?

First, scientific experiments have shown that just three years of a maize/legume intercrop can increase the infiltration of rainwater by over 3.5 times.[58] This is not at all surprising. In the African Sahel, the average SOM level in smallholder fields is less than 1%. I have often gone out into such fields right after a three- to four-hour downpour, and dug down into the soil. The water from all that rain has typically moistened the soil to a depth of less than 5 cm. The possibility that the infiltration rate in such soils could be multiplied by a factor of four or five is rather obvious.

SOM also increases the water-holding capacity of the soil. But this increase probably reaches less than 20%, even with heavy applications of OM. By itself, this small increase of water-holding capacity could not explain the degree of drought-resistance the Dogons were enjoying.

However, something else happens when good gm/ccs are used year after year: the *amount* of topsoil increases. At an average rate of increase in the topsoil of about 2 cm/year, even the Dogons, some of whom had only started using these techniques just over twenty years ago, could easily have accumulated 20 cm of additional topsoil. If that extra topsoil has even a 5% level of SOM, the total water-holding capacity of the A horizon could have doubled. Soil analyses have confirmed this increase in SOM.[59] The most dramatic case is that of the 3rd maize/mucuna system in northern Honduras. Soil analyses there showed a difference in SOM of between just 1 to 2% in surrounding fields and 12% where farmers had been using mucuna.[60] Mucuna is, admittedly, one of the best gm/ccs for producing large amounts of biomass, but the Dogon system has several gm/ccs (the trees plus leguminous crops) covering much of the soil, so it may well produce similar amounts of total biomass.

Once this moisture has been stored in the soil, the trees help keep it there by reducing the evaporation and transpiration rates. Soil moisture can then serve more to help crops grow, rather than being wasted in cooling the environment.

Lastly, the shade from a mulch also reduces soil moisture losses. Most home gardeners are very aware that a dead mulch will greatly increase the time that their soil stays moist after a good rain. Green mulches—the shade produced by living cover crops—will accomplish much the same thing. Some agronomists argue that since green mulches are also transpiring water, they do not maintain soil moisture any better than weeds. But farmer experiments have shown the contrary. Soil does stay moist much longer when a few plants that grow horizontally with broad leaves are covering all the soil, as opposed to perhaps ten times more weeds that grow vertically, with smaller or narrower leaves that allow a lot of sunlight through. Furthermore, green mulches naturally cool the environment because of the endothermic nature of the photosynthetic process, thereby reducing evaporation and transpiration rates beneath them.

If one combines all the effects of 1) a tripling of rainwater infiltration, 2) the added water-holding capacity of additional topsoil, 3) an increase in water-holding capacity of the soil in general, 4) the added infiltration of water down the holes made by termites and earthworms coming to the surface to consume fresh OM, 5) the decrease in evaporation and transpiration rates caused by the mulching effect of both the dead mulch and the live mulch, 6) the shade of the dispersed trees and 7) the lowered wind velocity caused by the trees, which also reduces evaporation and transpiration rates, one can begin to understand why gm/ccs can have so much impact on soil moisture retention. The impact shown in the two photos from Madagascar at the beginning of this book no longer seems so hard to believe.

In retrospect, this resilience to drought should not be too surprising. Most agronomists who spend much time in drought-prone areas have had the experience of walking from a bone-dry field suffering from drought, into a neighboring forest. In the forest, the soil is often still moist to the touch, even on the surface, and the natural vegetation shows no impact at all from the drought. Basically, by having a wide variety of plants, maximizing their production of biomass, keeping the soil covered, providing shade and having plenty of OM in the soil, the forest has overcome the drought very nicely. By closely imitating the conditions that have kept these forests alive for millions of years, farmers can create similar conditions in their fields.

That is the theory, but can it work in practice? Use of gm/ccs to increase resilience to drought, though still uncommon, is not limited to the Dogon people. I have personally seen the fields of farmers who have achieved amazing drought-resistance in at least six nations: Honduras, Mali, Zambia, Malawi, Mozambique and Madagascar.

The Zambian case is particularly interesting. Near Chipata, in eastern Zambia, an excellent NGO called COMACO has been training farmers to grow gliricidia bushes in rows in their fields, pruning them close to the ground several times a year in an alley cropping system (S48). Together with personnel from another NGO, I visited several of the more than 1,000 farmers who had been practicing this system for six or seven years. Typically, the COMACO farmers' 7-foot-tall maize was producing somewhere around 3 t/ha. The maize plants in neighboring fields, which had been planted the same week, were knee-high, with their drought-ravaged leaves curled up tightly.

One of the agronomists with me asked one of the COMACO farmers, "What is happening here? Why is there such a huge difference between your field and those of your neighbors?"

With a subtle, self-assured smile, the farmer replied, "Well, those of us who have planted gliricidia have prohibited the droughts!"

And so they had.

In this case, as well as similar cases in Madagascar, Mozambique and Honduras, the impact had been achieved without even any shade; the soil cover, both green and dry, and the increase in SOM, were the only factors that had created these dramatic differences. Furthermore, in each situation the farmers had used a different species of gm/cc. In Mali and Honduras they used multiple species, whereas in Zambia they had pruned gliricidia as a bush, in Mozambique they used jack beans or lablab beans, and in Madagascar a 9-month cowpea.

Scientists in temperate climates are also discovering the same impact of biomass on water retention. Dr. Steven Apfelbaum, the chairman of Applied Ecological Services, says research has shown that the flow of water off agricultural land has increased by up to 250 times what it was before the advent of agriculture. He summarizes his work on thousands of sq km of Minnesota floodplain by commenting that where biomass and soil cover have been maximized, "landscapes are hanging onto water and soil.…We're restoring nature's stinginess."[61]

How to "Prohibit the Droughts"

By never plowing their soil and keeping the soil covered, the Dogon people were clearly illustrating the four main principles of prohibiting droughts:

1) Keep disturbance of the soil to a minimum;

2) Keep the soil well-covered with live or dead mulches, or both;

3) Protect the fields from the tropical sun using dispersed or intermittent shade; and

4) Most importantly, increase significantly the OM content of the soil.

These principles are simple and self-explanatory, and farmers can put them into practice very cheaply. In fact, the cover can reduce the labor needed for weeding, and the dispersed trees usually provide firewood right near the home, so no one has to go far to collect it. Many of these systems thereby actually reduce farmers' total expenses.

In the beginning, while droughts can still have a major effect on crops, one should use gm/ccs that can withstand droughts. The most drought-resistant, fast-growing gm/ccs we have to date are the jack bean, the ratooned pigeon pea, and, after the first three months of its growth, the lablab bean (see Chapter 6). For truly drought-stricken situations, any of the *Piliostigma* species is probably the most drought-resistant of all, but it grows slowly the first couple of years, and we still need to learn more about how to manage it.

Despite its remarkable potential, prohibiting droughts still has several hurdles to overcome. It takes three to four years to achieve a noticeable impact; completely prohibiting droughts will not occur within two or three years; farmers may have difficulty believing that prohibiting droughts is even possible in the first place; and severe droughts during the early years may cause the gm/ccs to grow poorly, making the process even slower than originally hoped. Therefore, farmers may begin to seriously doubt the impact these practices can have.

By far the best way to overcome this skepticism is to take some of the farmer leaders on educational field trips to visit farmers who have already been successful at prohibiting the droughts, if such an experience exists anywhere nearby.

Another possible way to provide evidence is to pick a very small patch of land where some trees already exist, prune them well, and artificially add a large quantity of compost to the soil (about 100 kg of fresh compost/sq m of land), mixing the compost well with the top 20 cm of soil. Plant the crops in as similar a manner as possible to that used in the area, but using soil cover and gm/ccs. The results should definitely surprise local farmers. Such a demonstration should be located a good distance from any water source, or along a well-used path. Otherwise, people may just decide that the plot must have been secretly irrigated.

Chapter 6. Green Manure/Cover Crop Species

A Note about Scientific Names

Scientific names were originally established so each species of plant or animal would have only one name throughout the world. Yet over time, the names of many gm/cc species have been changed repeatedly, thereby defeating to some extent the original purpose of the system. Lablab beans have, at different times, been called *Dolichos lablab*, *Dolichos purpureus*, *Dolichos bengalensis*, *Lablab purpureus* and *Lablab niger*. Mucuna has been called *Stizolobium pruriens*, *Stizolobium aterrium*, *Stizolobium niveum*, *Stizolobium deeringiana*, *Mucuna deeringiana*, *Mucuna pruriens* and *Mucuna utilis*. Some of these names officially refer to only one species of mucuna, but each of them has at times been used to refer to all mucuna species.

In this book, the first time I mention each species, and once again in Annex 3, I give the scientific name that in my experience is the most commonly used around the world. However, in this chapter, where I describe each species I will list most of the scientific names that have been used for that particular species.

Differentiating Between the Seeds of Different Gm/cc Species

When distinguishing between various species or genera of legumes, color, shape and size are not nearly as important as is the shape, color and position of the hilum, which is where the seed was attached to the pod. For instance, lablab beans come in half a dozen or so colors, but all lablab beans have a very characteristic bright white, raised hilum that traverses more than half of one side of the seed (Photo 12). Mucuna seeds also come in a bewildering variety of shapes, sizes and colors (including cream-colored, black, brown and speckled), but they always have an almost-white, raised, elongated loop on the hilum that rather easily detaches itself from the seed.

Some gm/cc seeds are quite easy to identify. For instance, all seeds from the genus *Canavalia* have a similar long, grayish hilum (see photo 15). If the seeds are white, they are jack beans; if they are red, they are sword beans (*C. gladiatus*), etc. They vary considerably in size, up to a good 3 cm in length (easily some of the world's largest beans). At first sight, this hilum might be confused with that of fava beans, but the hilum of the fava bean is on one end of the seed rather than along the side.

The Role of Perennials

In the May 2016 *Scientific American* article,[62] CGIAR personnel presented perennials as a class by themselves, uniquely capable of improving soils. In fact, perennials do have an advantage over annual gm/ccs in two specific situations. One is in areas plagued by droughts. Annuals are particularly susceptible to droughts because every year they have to grow a completely new root system. While these roots are still tiny, like short strands of hair, the plants will obviously be very susceptible to drought. Perennials, on the other hand, may have hundreds of years of root growth behind them when a drought hits. Just like forest trees, perennials can, after the first year, withstand even rather prolonged spells of drought.

In the semi-arid to arid African Sahel, farmers tend to favor perennial gm/ccs like gliricidia, *Faidherbia* (*Faidherbia albida*) and *Piliostigma*, rather than annuals. In such dry climates, we should also usually treat lablab beans and pigeon peas as the perennials they actually are, by cutting them off (ie ratooning them) before the next rainy season. Lablab beans should be cut off at ground level, while pigeon peas should be cut off at about 40 cm in height. They will then re-sprout, taking advantage of their previous years' root systems to access water.

Perennials also have an important advantage in unshaded lowland areas with unimodal rainfall patterns, which make up a large portion of the tropics. In these areas, most annuals will die before the rainy season ends. Then, for anywhere from five to eight months, their dead plant residues lie in the sun, suffering from volatilization and heat. They will be useless as gm/ccs by the time the next crops are planted. It is precisely this factor that explains why so many experiments with annuals like cowpeas, peanuts, soybeans, and mucuna have failed to improve soil fertility.

Normal short-term annuals can only be used as gm/ccs in these areas if there is either enough residual moisture in the soil or enough scattered rain during the dry season that the annuals can be relayed into the main crops or planted after the main crops have been harvested. When these species are grown through the dry season, volatilization and synchronicity will no longer be problems (see S36). Other possible uses of annual gm/ccs in these situations include short-term annuals like 60-day cowpeas (that can often fertilize the same crop with which it is grown) or long-term annuals like the Lojy Be cowpea (that can survive until the beginning of the next rainy season).

Nevertheless, for areas above 1,500 m in elevation, for lowland areas with a bimodal rain pattern, or for areas with either a natural or man-made dispersed or intermittent shade, we can use dozens of annual species, such as cowpeas, mung beans, rice beans, soybeans, fava beans, chickpeas, annual mucunas, etc., that improve soils just as well as perennials do, or better. They also often provide other advantages the perennials cannot (e.g. very few perennial gm/ccs provide edible food), and annuals do not take years to produce seeds or become full-grown, as trees do. Any one of these advantages of annuals can become a critical factor in a variety of specific situations.

Limiting ourselves to only perennials would be like trying to walk with only one leg.

The Major Lowland Species

The lowland species, for areas between 0 to 1,500 m in elevation, are listed below, starting with the tropical world's most important species, and continuing roughly in order of descending importance. My judgement as to a species' importance is based on its present cultivated area, the number of people who consume it, and its potential for future expansion among smallholder farmers.

1. Multiple species

Many people seem to assume that only one gm/cc species can be used at a time in a given farming system. While it is true that gm/cc species are usually best introduced among smallholder farmers one or two at a time, the world's most successful gm/cc systems often include anywhere from 3 to 10 different gm/cc species.

The farmer-developed and still evolving Dogon system near Koro, Mali (S69), very likely stands as the most interesting example of such a system. Between its intermittent shade trees and leguminous crops, it includes some

7 to 10 species of gm/cc. It is also the most dramatic case I have ever seen of semi-desert conditions being improved not only in their drought-resistance, soil fertility, biodiversity and the productivity of food, but also in their overall environmental health. That maximum biodiversity and all these other important benefits exist in the same system is probably not a coincidence.

Whether most smallholder gm/cc systems will ever achieve this level of biodiversity is yet to be seen. Some Brazilians are working with as many as 8 or 10 different species in a system. The most notable of these is Ademir Calegari of IAPAR, although his work is mostly with medium- to large-scale producers. Nevertheless, virtually every one of today's smallholder systems could be enhanced by the addition of more gm/cc species. 90% of the world's lowland systems could probably profit from the addition of dispersed or intermittent shade. A large majority of systems would likely benefit from the intercropping of 60-day cowpeas. Millions of others could profit from the addition of crawling cowpeas, such as the Lojy Be.

Other things being equal, more biodiversity is almost always beneficial. A good mixture of several gm/ccs provides a better and longer-lasting cover, a more balanced release of nutrients, greater weed suppression, a greater diversity of soil microorganisms, reduced risk, and a more diversified diet. Furthermore, where good soil cover has been achieved by hauling in mulch material, additional species could save huge amounts of labor and allow the crops to feed from the mulch.

On the other hand, certain conditions limit gm/cc biodiversity. Competition for water between different species, especially before farmers have improved the infiltration and water-holding capacity of their soils, can prevent the use of more species. Lack of access to seeds of many gm/cc species limits the biodiversity for most smallholder farmers. Free-grazing animals also limit the number of species many farmers can use. And last but far from least, the lack of knowledge of extensionists about how to add more gm/ccs to farmers' existing systems keeps farmers from maximizing their biodiversity.

2. Lablab beans

Lablab beans (*Dolichos lablab*, also *Lablab purpureus*, *Dolichos purpureus*, *Dolichos bengalensis*, and *Lablab niger*) are very likely second only to pigeon peas as the most widely planted and consumed of all the world's gm/ccs. Lablab beans richly deserve this position because they fertilize the soil extremely well, control weeds well, produce around 50 t/ha/year of biomass, can easily be intercropped with maize, and resist drought very well after the first four months of growth. Furthermore, the dry beans, green

Photo 12. Honduras. These lablab beans are still green. Dried, they may be off-white, red, black, brown or speckled.

The pods are usually the shape and size shown, but very occasionally are larger. The pods generally grow up above the rest of the plant, making the harvest particularly easy.

In some places, rumor has it that lablab beans are not edible. In fact, they are eaten by hundreds of millions of people in India, and millions more in Kenya, Tanzania, Peru, and several Asian nations.

The beans contain 23% protein, which is about 50% more than common beans.

peas, pods, leaves, and tender flowers are all edible and tasty when well-prepared. On top of all that, the whole plant is one of the most nutritious and palatable of fodders for animals—one that can be available, green, all year round. Thus, the lablab bean, like several other gm/cc species, can totally eliminate any need for making hay or silage in the lowland tropics.

I have personally found lablab beans being widely consumed in northern Peru (as green peas), Ethiopia, Kenya (called "black beans"), Uganda, Rwanda, Tanzania, Mozambique (as dry beans), Malawi (as dry beans), India (as a pulse called "horsepeas"), Bangladesh (called "hyacinth beans"), Cambodia, Vietnam (as dry beans or tender green pods), Indonesia (as tender pods) and the Philippines (also as tender pods). Both the immature green peas and the dry beans have a slight bitter taste, but it can easily be masked by mixing them with peanut sauce, rice or some other food. Some varieties of the dry lablab beans must be cooked about an hour longer than common beans. As can be done with almost any high-protein food, lablab bean preparations are sometimes traditionally fed to malnourished women to stimulate breast milk production.

People also eat cooked lablab leaves in parts of Latin America and Africa, sometimes preferring them over cowpea leaves. In Zambia, people quickly blanch either the cowpea or the lablab leaves, then roll them up into balls the size of large marbles. They lay these balls out in a light shade for two or three days and then store them in old fertilizer sacks as a ready, virtually cost-free, high-protein food to be boiled and eaten any time the following

12 months. This practice could greatly benefit protein-malnourished people anywhere lablab beans grow. Rolling the leaves up before drying them has two advantages: first, about twice as many leaves fit into a given storage space, and second, the leaves in the center of each ball stay slightly green when cooked, producing a more appetizing-looking broth.

Lablab beans are especially valuable in drought-prone areas. Although quite susceptible to drought during the first three to four months after being planted from seed, they can subsequently grow and produce grain for up to six months with no rain at all. This means they can provide fresh beans up to six months after the harvest of almost all other field beans has ended—a very important consideration for families that experience a hunger season.

Lablab beans can also be ratooned, but we have not yet tested whether this means they retain after ratooning the tremendous drought-resistance they show after four or five months of growth from seed. If they do, lablab beans could rank among our most drought-resistant gm/ccs after their first year of growth. Furthermore, since the lablab is a long-term perennial (I have seen 25-year-old, woody-stemmed lablab trees in Haiti), farmers might be able to ratoon them for 30 or 40 years. These possibilities urgently need to be researched.

It could be that ratooned lablab beans might grow much faster than lablab beans that grow from seed, smothering the maize with which they are intercropped. In this case, farmers would have to prune the ratooned lablab a month or so after the maize is planted. Farmers are often reluctant to prune a food crop, but doing so would have two advantages: 1) it would be done early enough in the season that the N (from both the pruned biomass and the N sloughed off by the nodules because of the pruning) would serve as a side dressing for the maize, and 2) farmers in semi-arid areas could have an almost totally drought-resistant, high-protein, 6-month food source that would, in time, make their soils more drought-resistant—a series of benefits devoutly to be desired.

Lablab beans can produce 50 t/ha/year of OM (green weight). At that rate, if grazing animals do not consume a large part of the OM, the lablab could prohibit the droughts within four to six years and could increase and then maintain soil fertility for decades.

Lablab plants produce more beans and more over-all biomass when they can climb, so farmers almost always intercrop them with maize. Since they grow quite slowly their first few months, they only shade out extremely short varieties of maize. When intercropped, they should usually be planted at the same time as the maize, at about 3 seeds/sq m. If weed control is particularly important, 4 seeds/sq m may be better. Unless one can obtain bushy types, lablab beans should never be intercropped with sorghum or millet because the stems of these crops are too weak to support the lablab's weight.

Lablab beans are tremendously palatable for animals. Cattle prefer them to virtually any other fodder. In a village where cattle graze freely during the dry season and only one farmer has planted lablab beans, the village's cattle will all gather in that plot as soon as the dry season starts and finish it off before eating anything else. However, for farmers who have year-round control over their land, lablab beans can provide a wonderful year-round fodder (see S17). The plant as a whole contains about 21% protein (see photo 13).

Photo 13. Honduras. These lablab plants, already 3 months into the dry season, will stay green and produce seeds all through the rest of a 6-month dry season, thereby providing human food and highly nutritious, protein-rich fodder for cattle right when both people and cattle have the least to eat.

The main challenge with lablab beans is that they do not grow well in degraded soils, unless cowpeas, pigeon peas, peanuts or lablab beans have grown there the previous year to increase the number of rhizobia.[63] The best way to prepare a poor soil for a lablab crop is to intercrop bushy-type jack beans with maize, sorghum or millet the year before planting the lablab.

Another challenge for lablab beans is insect pests. Since people usually want to eat the beans, some form of pest control will often be necessary.

In moist climates, aphids can be especially troublesome. A lack of moisture kills aphids, so one can often control them by spraying the lablab pods with a powder made of cassava or wheat flour or something else that will dry out the surface of the pods.

3. Cowpeas

Cowpeas (*Vigna unguiculata,* also *V. sinensis*, *V. catjang*, *Dolichos sinensis*, *D. unguiculatus,* and *D. biflorus*) are another of the most widely consumed legumes in the developing world. The most common cowpea varieties grow as small bushes that mature after 90 to 120 days. These are frequently intercropped with maize, sorghum, millet or cassava. Cowpeas fix about twice as much N as common beans and are much more tolerant of heat, drought, and soil acidity. Nevertheless, in lowland, unimodal areas without shade, they have minimal value as gm/ccs because they produce little biomass and lose virtually all their N before the following growing season.

Cowpea leaves are widely consumed in Africa and Asia, as well as parts of Honduras, Nicaragua and El Salvador. They can also be rolled up and dried, so this virtually free high-protein food can be enjoyed throughout the year (see the section above on dried lablab bean leaves).

The young cowpea pods are also eaten across much of the world. These pods look much like western string beans, except that cowpea pods are narrower and more rounded. The meter-long bean (*V. sesquipedalis*), primarily consumed as a green pod and found in much of eastern Asia, is closely related to the cowpea.

We have recently found that three little-known varieties of cowpeas—the crawling cowpea, the crawling 9-month cowpeas (Lojy Be) and the 60-day determinate cowpeas—may have tremendous potential as gm/ccs.

Crawling cowpeas. Crawling cowpea varieties are unknown in most of the world, but are used widely in Mozambique and Malawi, as well as a few villages across the Sahel, Laos and Thailand. The crawling cowpea's major advantage is that it does not climb other crops, but still provides a fair amount of biomass, N, soil cover, weed control and high-protein grain. It is therefore used traditionally as an intercrop with maize, sorghum and cassava, but also has tremendous potential as an intercrop with vegetables (see S28) and with upland rice, for both of which we have had a hard time finding good gm/ccs.

The crawling cowpea could also be used with systems that already have intercropped legumes but allow a fair amount of light to reach the soil, such

Photo 14. Indonesia. Although this photo was taken in Indonesia, cowpeas are intercropped with maize the same way throughout most of Africa and much of Central America.

In Africa, cowpea leaves are often the first fresh food available after the rains start, thereby helping to alleviate the hunger season. The grain usually starts producing about two months later, which often marks the end of the hunger season. If farmers can get seeds of a 60-day cowpea, they can cut the hunger season by a whole month!

This possibility is especially appreciated by Africa's women, since they are doing most of the weeding precisely when they are hungriest.

as maize and pigeon peas or cassava and pigeon peas. Such a system, called a "doubled-up legume," is already being promoted in Malawi, using pigeon peas together with peanuts or soybeans (see S21).

When intercropping crawling cowpeas with maize, the best time to plant the crawling cowpea seems to be one to three weeks after the maize is planted.[64]

The Lojy Be. The Lojy Be (pronounced LOO dzeh BEH), a very recent discovery in southwestern Madagascar, may turn out to be a dream come true. It is a crawling cowpea that yields grain over a period of four months, produces a good deal of biomass, and has a total life cycle of nine months.

For years in Africa we have been looking for a good gm/cc that can dramatically improve soils and that is also widely eaten or is marketable at a good price. Lablab beans improve the soil and are marketable in Kenya and neighboring countries but are unknown and often unappreciated elsewhere. Pigeon peas have also had a great role to play in much of eastern Africa the last few years because the Indian market was paying very good prices for them, but the price has since crashed, and in much of Africa, pigeon peas are not appreciated as a food. Peanuts and, to a lesser extent, mung beans, are widely consumed, but are poor gm/ccs. A legume that could fulfill both roles almost anywhere in the lowlands of Africa would greatly speed up the adoption of gm/ccs across the continent.

The Lojy Be accomplishes both objectives and much more:

1. Cowpeas are among the three most widely consumed grain legumes in Africa, and are consumed in large quantities—enough to dramatically improve people's nutritional well-being if they had access to them most of the year.

2. Most cowpea varieties fix about 60 kg of N/ha, but in unshaded, unimodal environments, are largely useless as a gm/cc because they die after less than 5 months, so their N is volatilized during the dry season. By contrast, the Lojy Be variety can last throughout the whole dry season after anything but the driest rainy seasons, making it a very good gm/cc. That is, it can gradually increase yields over a five-year period to two or three times previous yields and, in the process, give the soil an astonishingly high level of resilience to drought (see photos 3 and 4).

3. The Lojy Be variety is more resistant to drought than most other cowpeas.

4. The Lojy Be variety produces grain for a period of four months, from about three to seven months after planting. Then it normally stays alive another two whole months. As a result, the following season's maize, sorghum or cassava will receive a full supply of N and other nutrients precisely when they most need them.

5. As with all cowpea varieties, the leaves of the Lojy Be can also be eaten or dried and stored for use throughout the following 12 months.

6. Lojy Be plants cover the ground almost completely from about 40 days after planting until they die. The result is that soils have more drought-resistance and are undamaged by the tropical heat.

7. Being a crawling cowpea, the Lojy Be can be intercropped with virtually any other crops already planted in farmers' fields, except other crawling plants or cover crops.

8. At least in Madagascar, people consider the Lojy Be to be tastier than most other cowpeas.

Of course, no crop is perfect. As with all cowpeas, the Lojy Be is attacked by several insects that must usually be controlled if the cowpea grain is to produce well. With grain being produced over a period of four months, much of the grain will be maturing when no other cowpea grain is around, so the insects may all converge on the Lojy Be (until other farmers also adopt Lojy Be). Furthermore, if the rainy season is excessively droughty and there are no rains at all during the dry season, the Lojy Be will occasionally die prematurely.

These are, however, minor concerns. The farmers' seed supply will not be affected by the droughts because plenty of seed will have been harvested before the dry season starts. The insect pests can be controlled by already known natural or chemical means. And the susceptibility to droughts will gradually be overcome by dispersed shade or the gradual improvement of the soil's moisture retention brought about by the Lojy Be itself.

60-day cowpeas. The 60-day, determinate cowpea can also gradually raise yields throughout most of Africa, Asia and Latin America. Two characteristics of the 60-day cowpea make this possible. First, it produces dry beans ready to harvest within 60 days of being planted. Second, if it is planted immediately after a heavy rain, it will produce a fair harvest of beans, even though it receives no more rain at all.

If the 60-day cowpeas are planted at the same time as the other crops, near the start of the rainy season, they can shorten people's hunger season by a full month. This characteristic has already made 60-day cowpeas extremely popular in parts of Mali, Burkina Faso and Mozambique, especially among women.

Their short cycle also means that 60-day cowpeas can be intercropped with maize, sorghum or millet along with any other plants that are intercropped with these basic grains, as long as the other crops do not occupy all the space in the field during the first 50 days. For example, one can have maize and pigeon peas in a field, or cassava and jack beans, and still add 60-day cowpeas to the mix. By the time the other two crops have grown enough to need the space occupied by the cowpea, the cowpea has already been harvested. Farmers can thereby add a whole new crop to their already-intercropped fields.

Farmers can also turn the cowpea into an effective gm/cc. The cowpea biomass applied to the soil about 60 days after the main crops have been planted can act perfectly well as an N side-dressing to the main crop, thereby not only improving the yield of the main crop, but putting its N back into the farming system before the dry season problems destroy it.

Furthermore, the 60-day cowpea can be planted when the first rains come at the end of the dry season. Often farmers cannot plant anything after these first rains, because another month will often pass before the rainy season actually begins. But since the 60-day cowpea needs no rain after it is planted in a good moist soil, it will grow perfectly well during the month or so before the rainy season does start. In such a case, the N from the harvested cowpea will be added to the soil only 30 days after most of the crops are planted. Another tremendous benefit from such a circumstance is that the 60-day cowpea will be shortening the people's hunger season by two months instead of just one.

These two systems will have an advantage, unique among all our tropical gm/cc systems, except the *Sesbania rostrata* system (S68). The crop that is planted in the same season as the gm/cc can be fertilized by the gm/cc, so that recognizable results will occur within two or three months after planting, rather than 8 months later in bimodal areas or 14 months later in unimodal areas.

Farmers in Malawi and Zambia have found still another valuable use of the 60-day cowpea: growing them at the end of the rainy season. If there is a good, heavy rain toward the end of the season, or several lighter rains that leave the soil good and moist, the 60-day cowpea can be planted during these rains, and will then produce grain well into the dry season, even if it receives no more rain at all after being planted. In these ways, with a little luck as to rainfall patterns, many farmers will often be able to grow two crops of cowpeas during and after the same 6-month rainy season.

4. Jack beans and other species of Canavalia

The jack bean (*Canavalia ensiformis* and *Dolichos ensiformis*) with white seeds, the red-seeded sword bean (*C. gladiatus*) and other members of the genus Canavalia, are the most drought-resistant gm/ccs during their first year of growth of all the legumes we know (photo 16). They are also extremely resistant to poor soils, are affected by almost no pests, can fix an astounding 240 kg/ha/year of N, and produce over 40 t/ha of biomass (green weight). One member of the family, *C. maritima*, even grows on ocean beaches, with saltwater lapping around it at high tide.

These characteristics make the *Canavalia* species the best for recuperating wastelands. The bushy-type jack bean also shares first place with the pigeon pea for starting the process of increasing the fertility of badly degraded land that can still be farmed. Bushy-type jack beans can be intercropped very well with maize, sorghum, millet, or cassava to jump-start the soil restoration process without losing a year's production of subsistence crops. Occasionally these associated crops will suffer anything up to a 20% decrease in yields the first year, but will usually enjoy at least a 50% increase in yield by the second year. This is another good reason for always trying a new technology on a small scale: the potential 15% decrease in yield never needs to occur on all of a farmer's land at once.

Smallholder farmers often object to growing a gm/cc that produces no benefits other than more fertile soil, but if they know they only need do so for two or three years, they are usually quite happy to try it out. Therefore, after one to three years of using jack beans—

Photo 15. Haiti. Jack bean pods and seeds. In areas where firewood is extremely scarce, the large jack bean pods can be used as firewood. They contain an oily substance that makes them burn longer than other pods.

If you are unable to obtain jack bean seeds from normal commercial sources, they can often be found within most African countries and even some Latin American countries. Show village women a photo like this one, and they will be able to bring jack bean seeds from the forest. However, the jack beans in the forest tend to be climbing varieties, which are fine for soil recuperation, but cannot be used very easily for intercropping.

As far as we know, the jack bean is the only member of the *Canavalia* genus that has a bushy type that can be intercropped well to restore degraded land that is still being farmed.

once the land has become reasonably productive—most farmers switch to another gm/cc species that produces more benefits than the jack bean. Still, over time, the jack bean's ability to restore soils where nothing else can, usually wins it enthusiastic converts.

This ability of the jack bean to restore soil fertility and provide drought-resistance has been underestimated by almost everyone. In Nampula Province in northern Mozambique, where the sandy soils are so depleted that farmers reluctantly gave up growing maize long ago, farmers were producing well over 1 t/ha of maize after just one year of jack bean. In a nearby trial, 30% of the cassava plants had died of an early-season drought in the control plot with no previous jack bean, while in the test plots, following one year of a cassava/jack bean intercrop, the cassava achieved over a 95% survival rate and produced twice the above-ground biomass as in the control plot. Farmers commonly notice some resistance to drought in a field within a year or two of adopting jack beans.

Jack beans may become even more popular with the recent discovery that the 60-day cowpea can usually be intercropped with both the jack bean and a staple crop without occupying any additional space, because it is harvested before the other plants need the space. This discovery means that people

Photo 16. Mexico. In this drought-prone area of Yucatán, Mexico, most of the maize died of drought, but the jack beans show no sign of moisture stress. In Haiti, I have seen jack beans growing very well even when the sorghum with which they had been intercropped had all died of drought.

can harvest an edible bean from the system even though the jack bean itself produces no food.

The jack bean grows well in moderate shade, thereby doing well as an intercrop. After five months of wet season, it can also grow through most or all of a 6-month dry season, thereby avoiding the problem of N volatilization.

As mentioned earlier, jack beans do not do well in soils with a pH of less than 4.5, but they do grow well in other depleted soils. On the other hand, jack beans do not grow well in highly fertile soil or in areas of heavy rainfall.

For the most part, cattle do not eat jack beans, but in areas with an annual rainfall of less than 400 mm, they sometimes will do so toward the end of the dry season.

Sources differ, sometimes claiming that the thin, immature pod, and even the seed, are edible. However, I would recommend eating only the tender pod, but only as an emergency famine food. Just recently, there is news that the jack bean seeds can be eaten after a relatively short process of being boiled several times. Final testing of the process is still ongoing.

Two species of *Canavalia* face possible extinction. One, which ranges from northern Guinea into southern Mali, has purple-and-white filigree seeds that are very large (up to 3 cm in length) and easily among the most beautiful bean seeds in the world. The other species, from the area around Kolokani, north of Bamako in Mali, has seeds almost as large with a black-and-white filigree. Because of their size (being able to hold more moisture) and natural habitat, I suspect these seeds could be even more drought-resistant than the jack bean. At present, these seeds are mostly used for making necklaces and as tokens in board games. If anyone could collect and multiply these seeds, s/he just might provide a major service to farmers in drought-prone areas all over the world.

5. Pigeon peas

Surprisingly, pigeon peas (*Cajanus cajan*, also *C. indicus*, *C. flavus* and *Cytisus cajan*) are the fourth most widely consumed grain in the entire world, after rice, wheat and maize. Over a billion people in India regularly consume pigeon peas, as do millions in China, West Africa, and the Caribbean.

Pigeon peas grow on a 1- to 3-m tall bush. They grow well in areas with an annual rainfall as low as 500 mm and are quite resistant to drought, but may produce seedless pods when under serious water stress. Though the plant usually dies after four or five years, it produces its best grain yields during the first three years. Sometimes it grows rather poorly the first year in a given field, probably because of a lack of rhizobia, but after that, it can fix up to 200 kg of N/ha.[65]

The pigeon pea can be grown simply intercropped with maize, sorghum or millet, or it can be grown in doubled-up legume systems by adding peanuts, soybeans or 60-day cowpeas to these systems (see S21). Shorter-term varieties do exist, but the pigeon pea usually takes 6 to 8 months to produce grain. Thus, smallholder farmers can only grow most varieties of pigeon peas where livestock grazing is controlled during at least the beginning of the dry season.

Pigeon peas can also be ratooned, making them as drought-resistant as many trees. See S20 for a description of the process.

We have recently begun experimenting with another very promising new use for pigeon peas: as dispersed shade in multi-species kitchen gardens (see S6). Planted every 1.5 to 2.5 m square, they can both produce food and provide a 4-year dispersed shade for any vegetables that grow to less than 1.5 m in height.

The pigeon pea grain is tasty and highly nutritious, with a protein content of about 20%. The green pods and green peas can also be eaten in salads. The dry, uncooked grain makes an excellent feed for most animals, but needs to be cooked before being fed to fowl.[66]

6. Mucuna or velvet bean

Mucuna (*Stizolobium pruriens*, also *Stizolobium aterrium, Stizolobium niveum*, *Stizolobium deeringiana*, *Mucuna deeringiana*, *Mucuna pruriens* and *Mucuna utilis*) is commonly also known as velvet bean, as well as many other names in different locations. Mucuna has for years been considered the most important of all the gm/ccs. It has, after all, been grown for decades by hundreds of thousands of smallholder farmers in around two dozen nations in Latin America, Africa and Asia. Nevertheless, because of its prominence, it has been promoted in many situations where we now know other species are much more useful. As a result, to the extent that we have learned more about these other species, mucuna has declined in relative importance, and will probably continue to do so.

Mucuna does, however, continue to be popular among farmers in certain situations, despite its not being easily edible (aside from being a coffee substitute). This popularity undoubtedly results from the climbing varieties' being by far the best gm/cc we have for controlling weeds. That farmers use

Photo 17. Honduras. The mucuna used as a gm/cc, on the right, must never be confused with the itchy kind of mucuna, to the left, the pods of which are covered by small, extremely irritating, fuzzy hairs. Notice that the farmer here is only touching the normal mucuna.

mucuna so widely provides an important indication of smallholder farmers' priorities: reducing the demand for labor is high on the list.

Mucuna's superior ability to control weeds comes from a combination of three factors:

1) Mucuna is extremely aggressive. Mucuna, along with lablab beans and runner beans, produces an incredible amount of biomass very quickly, basically outgrowing its competition and shading it out. It is so aggressive that when people first see it crawling over everything in sight, they often fear it will become a pest, killing everything in its wake. I have seen cases where it had covered and killed full-grown orange trees and grew to the top of 20 m tall pine trees. Yet I have never heard smallholder farmers complain about the annual mucuna becoming a pest. This is likely because farmers can eliminate it very easily. If it is cut off once, below the first branches and before it has produced viable seed, all but a very few plants will disappear from a field. (The much less common perennial mucuna, has caused problems by retarding forest regeneration in a few areas.)

2) Mucuna also has the uncanny ability to place its next leaf precisely where it will best prevent any light from penetrating its canopy.

3) Mucuna's ability to control weeds may also come from allelopathic substances (probably L-Dopa) in its leaves that deter weed growth.

In addition to controlling weeds in general, mucuna can get rid of several particularly noxious weeds and parasites. Specific methods for eliminating imperata grass, nut grass, and wild sugarcane are described in text boxes S11, S13 and S14, respectively.

In rare cases, mucuna may fail to control local weeds. For example, on Kalimantan Island in Indonesia, some very aggressive native weeds outgrow mucuna. Also, mucuna may not grow enough to control weeds where wild animals, like deer, feed on it.

S1. Any crop/itchy mucuna. NEVER. Anywhere. Some types of mucuna have a heavy coat of small hairs on the pods (see photo 17). These hairs are extremely itchy. They have earned these varieties of mucuna a number of negative names from "pica-pica" (itch-itch) in Central America to the well-earned "devil bean" in Nigeria. If a person touches them, they sting like nettles, but the pain lasts about three days. Affected people often scratch themselves until they draw blood. You do not even have to touch the pods to be affected. When the pods dry out (usually a month or so after the rainy season ends), these hairs can blow as far as 50 to 100 m in the air. This itchy mucuna should absolutely never be planted in farmers' fields, nor should anyone get within 100 m of it when the pods are dry.

These itchy mucuna varieties exist in forests throughout much of the tropical world. One variety of mucuna in southeastern Rwanda and across the border in northwestern Tanzania has the typical hairs but does not itch very badly. Virtually everywhere else the itching caused by the fuzz on the pods is intense.

The seeds of the itchy mucuna are always black, although the seeds of the non-itchy mucunas, which come in a variety of colors, are also sometimes black. In areas where the itchy mucuna exists, farmers should avoid using any black mucuna seed, even of the good kind, so they will not accidentally get seeds of the itchy kind mixed in with their own seed stock.

Mucuna is very photoperiod sensitive. It almost always produces seeds and dies during the shortest days of the year. If, for instance, it is planted anywhere from January through August in the northern hemisphere, it will die in December. However, if planted in October or November, it will continue to grow for more than a year, until the following December.

Mucuna fixes up to 140 kg of N/ha/year (the equivalent of about 6 sacks of urea) and produces up to about 60 t/ha/year of biomass (green weight). This OM often decomposes quickly, as does most leaf matter with a high N content. If the leaves are buried in a moist, lowland tropical soil without other crop residues, they tend to disappear within two to three weeks. However, in the tropics we generally want to slow down decomposition, so mucuna residues are best left on top of the soil. To the extent that they are mixed with OM from associated grain crops or other grasses with a lower N content, these leaves will take from one to three months to decompose, providing better soil cover in the dry season and making more of the nutrients available for the next crop. This process also largely eliminates any possibility of the groundwater being polluted by excessive N.

Of course, leaving the crop residues on the surface means that more N will volatilize. In unimodal lowland areas with no shade, perennial mucuna or some other gm/cc species would be a better option than annual mucuna.

Generally, trying to intercrop mucuna with any other crop is tricky, precisely because of its aggressiveness. Farmers find it difficult to keep the mucuna from smothering the associated crops. Farmers can plant the mucuna 20 to 40 days after the maize, but if they plant it too early, it will suffocate the maize. If they plant it too late, the mucuna will be stunted by the excessive shade from the maize. The best timing varies according to how tall the maize grows and how soon it matures. And the best timing may vary from year to year, depending on whether the rains allow the maize to grow well. Of course, mucuna can be pruned, but pruning requires a large amount of extra labor precisely when farmers are busiest. Furthermore, if it is pruned too low or too late, it will be stunted because of the shade of the maize. If pruned too early, it may require a second pruning.

If farmers still want to intercrop mucuna with their maize, they should try it out in small-scale plots until they find a management system that works for them. One possibility consists of planting the mucuna up to 20 days after the maize, and then pruning it down to half its height once it starts covering the maize. This heavy pruning will fertilize the maize with N that the legume releases, or "sloughs off." Another option is to plant the mucuna 30 or more days after the maize. Then, when the maize starts shading the mucuna too much, pull the lower leaves off the maize to give the mucuna more sunlight.[67]

Farmers can also relay mucuna into maize a month or so before the maize harvest, once again pulling the lower, almost dead leaves off the maize. In this case, the mucuna's biomass production may not be optimum because of a lack of rain, but in high rainfall areas or where the maize matures early, this late planting may cause the photoperiod-sensitive mucuna to stay green throughout a 6-month dry season.

Very likely the best option for intercropping mucuna with maize, sorghum or millet is that of using a bushy-type mucuna. The bushy-type mucuna from South America often produces too little seed, some of which rots because it is touching the ground, but a bushy-type variety introduced into Ghana by Bern Guri (guribern@gmail.com) produces plenty of seed. As a result, it has become quite popular with the area's farmers. More often, lablab, pigeon pea or some other legume would be preferable for intercropping with maize.

Under tree crops, mucuna is rarely the gm/cc of choice, because it must constantly be cut away from the base of the trees. For this purpose, perennial peanuts (see S85 and photo 43) or centrosema (*Centrosema pubescens*) would be preferable. Rare exceptions to this rule are described in S87 and S89.

Probably the best general use of mucuna is either in a very high rainfall area using the 3rd maize/mucuna system (S52), or as an improved fallow. Often, as an improved fallow, it can restore soil fertility as well in just six months as a natural fallow can in five to six years. Such an improved fallow can be used easily in a bimodal area. In a unimodal area, farmers should use a perennial mucuna, so it will grow through the relatively dry season.

In very drought-prone areas where mucuna might not grow well, farmers can reduce their risk by planting both mucuna and jack beans. In a good year, the mucuna will overcome the jack beans, but in a year too dry for mucuna, the jack beans will still grow well and provide good soil cover.

Human consumption of mucuna is somewhat controversial because it contains a number of antinutritional factors. Seemingly most alarming, it contains cyanide, but only "far below the lethal level."[68] Nevertheless, it is generally considered to be perfectly safe if properly treated. After all, a good number of foods we consume every day, such as cassava and many kinds of beans, contain traces of anti-nutritional factors.

On the positive side, mucuna's grain has a protein content ranging from 20 to 30%, with a quality of protein similar to that of soybeans. This means that the grain can provide a very inexpensive source of abundant, high-quality protein, ideal for people with protein malnutrition. Mucuna also has medicinal qualities. Its most common use in Western medicine is for extracting L-Dopa, the medicine of choice for treating Parkinson's Disease.

Above all, mucuna has been consumed by humans for centuries, and in large parts of the world. Mucuna beans are consumed in Ghana, Benin, Nigeria, Malawi, Mozambique, India, Sri Lanka, Indonesia, China and Japan. Millions of people in southern Mexico, Guatemala, Ghana, Kenya and elsewhere drink mucuna coffee as an inexpensive, caffeine-free coffee substitute.

Making a coffee substitute is the easiest way to prepare mucuna for human consumption. To make it, roast the mucuna beans until the insides are black, grind them and put the ground mucuna into a porous bag or a clean sock, which is then steeped in boiling water. The L-Dopa will remain in the coffee grounds. Mucuna coffee tastes enough like real coffee that if it is mixed half and half with real coffee, few people can taste the difference.

Photo 18. Honduras. Both the coffee and chocolate cake in this photo were made of mucuna. Even "chocolate" frosting can be made from mucuna with no use of cacao. This cake requires hours of processing, as described in the text.

For other uses, mucuna beans require major processing before being eaten. Traditionally, people boil the mucuna beans six times for several hours each time and throw out the water each time. Since this process requires such a large amount of both fuel and labor, a scientist at IITA developed a much more efficient way of making mucuna safe to eat. First, you crack the seeds, either by hand or preferably in a village mill at a very wide setting. Then you remove the seed coats and soak the seeds overnight. The next morning, you throw out the water, wash the seeds in fresh water and remove the remaining seed coats. Boil the seeds for 20 minutes, leaving them in the water until it cools. Throw away the water, wash the grains a second time and leave them to soak in more fresh water for a second night. On the third morning, wash the grains again in fresh water and then let them dry thoroughly in the sun. Finally, winnow the seeds to remove the last of the seed coats. Be sure to wear old clothes while doing all this, because mucuna juice stains cloth.[69]

7. Gliricidia or mother of cacao

Gliricidia (*Gliricidia sepium*), also called "mother of cacao," "quickstick" or "the rat killer tree," is rapidly gaining importance. Whereas leucaena has often been called the "miracle tree" for alkaline soils, gliricidia easily deserves the same name for acid soils. Every part of the tree has some use. The leaves rank as one of the best tree-based gm/ccs in the world, as well as being a highly appreciated dry-season fodder for cattle and the basic ingredient for a nutrient-rich foliar spray. The fast-growing branches provide large quantities of firewood, the bark provides a powerful rodenticide, the flowers are edible and, of course, the trunk sequesters carbon.

Not surprisingly, given all these benefits, the tree has taken on a whole series of diverse roles. Tens of thousands of kilometers of live fences of gliricidia stretch across northern Honduras, Guatemala, Madagascar and Southeast Asia. Gliricidia trees are considered the best tutors for vanilla and black pepper in countries as diverse as Mexico, Madagascar and Indonesia. The traditional "quesungual" system in southern Honduras from which the United Nations' Food and Agriculture Organization (FAO) developed the "dispersed shade" idea (S4), usually features gliricidia as dispersed trees. In Mali, gliricidia planted as a dispersed tree system has become the tree that saves people's cattle from starvation during droughts (S3). More recently, gliricidia-based alley cropping is spreading across parts of eastern Zambia (S48).

Despite its ability to thrive in high-rainfall areas, gliricidia is very drought-resistant. In northwestern Mali, farmers planted gliricidia seedlings in July, in depleted soil with no OM. Just 60 days later the rains stopped for a full 8-month dry season. Even so, 95% of the young trees survived that first year. It also does amazingly well in infertile soils. In the Kalahari sands of western Zambia (which consist of almost pure sand down to a depth of over 4 m), gliricidia planted as cuttings and ratooned pigeon peas with year-old root systems did better during a drought than did jack beans. In spite of their drought-tolerance, gliricidia trees should probably not be planted in areas with less than 400 mm of rain/year, as some of the trees have died in the northern Sahel.

As with all legumes, farmers should be careful to prevent bloating in cattle. Whenever cattle have been eating dry, non-leguminous material for some time, the amount of fresh leguminous material they eat should be increased gradually over a month or so. Gliricidia leaves should not be fed to non-ruminants at all.

A kilo of fresh green leaves pruned from a gliricidia tree can fertilize crops just as well as a kilo of fresh cow manure. Often, to motivate farmers to plant gliricidia, we bury a kg of fresh gliricidia leaves within a few cm of each of a couple dozen hills of recently planted maize or vegetables. The impact of the gliricidia on these crops almost always catches farmers by surprise and creates considerable enthusiasm for planting more trees.

Photo 19. Honduras. These gliricidia trees, used as a live fence and as a source of fodder and firewood, have been ratooned every year for about 20 years. These trees were planted as cuttings so they could serve as fence posts within two months after planting. A similar system of pruning is used for gliricidia in dispersed shade systems.

In the Sahel, by far the most popular benefit of the gliricidia-based dispersed tree system is that the trees start sprouting young leaves about 70 days before the rains come—precisely when, after a droughty year, people's cattle run out of fodder and start dying. At this time, farmers with a few gliricidia trees will have enough fodder to keep their cattle well-fed (see S3 and photo 42). If farmers need green fodder for their cattle throughout the dry season, they can prune all the trees' flowers so the trees will produce green leaves year-round.

In 2011, I helped establish, together with a group of women from a village called Debela, one plot of 50 gliricidia trees as dispersed shade. These trees are so popular that these women have spread the technology for up to 15 miles away from the village. There are now, seven years later, over 800 plots with a total of over 120,000 gliricidia trees! (If you wish to observe this phenomenon yourself, go to "Debela, Koutiala, Mali" on Google maps, move in as close as you can to the village, and then pull back one stop. The first plots people planted, with dark live fences around them, are visible near the village, but as you move out further, you will see smaller dots—younger trees—planted in a more or less rectangular fashion at roughly the same spacing, stretching for many km from Debela.)

Gliricidia leaves are not very palatable, so cattle unfamiliar with them may not start eating them until they are very hungry. Farmers in the Philippines often starve their animals for two or three days to get them to start eating gliricidia leaves. Once the animals have eaten some, they will easily eat them any time they lack tastier fodder.

For years, we had seen gliricidia leaves' lack of palatability as a serious disadvantage. But across the Sahel, and in other drought-prone areas, it turns out to be a blessing in disguise. Many people have claimed that gm/ccs would never work in drought-prone areas because farmers would always prefer to feed the gm/cc biomass to their cattle rather than to the soil. They are right—with one exception. Since gliricidia leaves do not taste very good, the cattle abandon them as soon as the rains come; they much prefer the tender new grass that is sprouting everywhere. By the time the crops need the OM—20 days to a month after the rains start—farmers are plenty happy to use all the new gliricidia leaves to fertilize their crops.

The flowers of the gliricidia are edible and are widely eaten in southern Honduras and eastern El Salvador, where the tree is native. In most of the world, they would probably be most accepted as an additional ingredient in the sauces commonly served with rice and other starches.

Farmers can use gliricidia bark to make a poison for rodents. In much of the developing world, rats and mice are problematic, in part because they eat a lot of food and seeds that farmers store in or near their homes. Rats often cause serious problems in rice fields, because the grain grows close enough to the ground that rats can reach it. To get rid of these vermin, one can put some of the food the rodents are eating in a pot, add a few strips of gliricidia bark, and boil them together for half an hour. Then the poisoned grain should be placed underneath a large rock or a heavy log, or inside a long piece of bamboo, where small rodents can get to it, but **where domestic animals and children CANNOT reach it.** As with most rodenticides, this poison can easily kill humans and other mammals. Gliricidia roots and seeds also contain some poison, but in a much, much lower concentration.

In areas where termites abound, be careful while pruning to leave at least four or five branches on the tree at any one time so enough of the trunk remains alive to keep the termites away (termites always prefer to eat wood that is dead).

Farmers can propagate gliricidia in three ways:

1) From seeds. If gliricidia trees are rare or non-existent in the area, large numbers of trees are needed, and there is plenty of rain, it is usually best to plant them from seeds.

   Gliricidia seeds must be collected when they are very dry if they are to germinate well. If farmers pick them directly off the tree, they will often have a germination rate of only about 30%. Farmers should spread an old piece of cloth or plastic under a tree, then shake its branches, collecting only those seeds that fall to the ground. Seed collected this way should have a germination rate of over 80%. Before planting, the seed must be soaked in warm to hot water for about three hours.

2) As seedlings. Making and managing nurseries is by far the most labor-intensive way of producing gliricidia trees. It also requires a reliable source of water precisely during the months when water tends to be scarcest in the villages. However, if you want to produce large numbers of trees in areas where no gliricidia trees exist, and if droughts threaten to kill trees getting started from seed, seedlings are the best way to go.

   To maximize gliricidia trees' survival in drought-prone areas, the seedlings should be:

   a) Transplanted when they reach a height of 50 cm above ground level. This means they will be larger than most seedlings when planted, so the nurseries need to be established about a month earlier than usual.

   b) Planted at the same time as people's crops.

c) Planted into planting stations deep enough so their taproots go straight down. If the seedlings are 50 cm tall, these planting stations will have to be fairly deep. But their increased survival rate, and the benefits these trees provide in drought-prone areas, will easily pay for the extra work.

d) Pruned of all except the five or six top branches.

e) Fertilized by burying the pruned gliricidia leaves near the base of the trees.

f) Protected by placing a small amount of mulch around the base of the seedlings.

For trees planted as seeds or seedlings, prune off all the lateral branches on the lower 2/3 of the stem at the beginning of the rainy season each year for the first three years.

3) As cuttings. In areas where gliricidia trees are quite common, cuttings are usually the best way to propagate them. Using cuttings eliminates all the work of establishing nurseries and gives the trees at least a year's head start. The cuttings should be at least 3 cm thick, but even thicker if you have a plentiful supply. They should generally be quite long (up to 1.5 m in length) so the resulting tree will have a long, straight stem and its leaves will be out of the reach of grazing animals as soon as possible.

In most soils, one should dig 50-cm deep planting holes. In soft earth, they can instead carve a point on the bottom end of the cutting and merely jab it into the ground. This method saves a lot of labor even if many of the cuttings have to be replaced.

In Africa, termites may infest cuttings during their first year because much of the wood in the new stem dies. If extra cuttings are available, one can just replace the affected cuttings. If cuttings are not easily available, find a natural way to control the termites, or make a one-time application of insecticide. Some farmers bury wood ashes near the trees' roots to keep the termites away. Another possibility would be to lay on the ground near the gliricidia tree a number of dead sticks of some species particularly favored by termites.

If cuttings are not available, but farmers like the idea of using cuttings and are willing to wait, plant a good number of gliricidia trees in the area as soon as possible, hopefully as dispersed shade in kitchen gardens that are being watered. Within two to three years, these trees will provide a good source of cuttings.

Two or three years after planting, when the top of the stem or the most vertical branch reaches a height of 2 m, the farmer should cut it off at a comfortable height just above his or her head. Each year thereafter, he or she should cut off most or all the branches a month before the rains, so the trees' branches will not shade the crops excessively (see photos 19 and 43).

Farmers can supplement dispersed trees with leguminous hedgerows planted along each second or third row of trees. Since dispersed trees are spaced far enough apart that they alone cannot increase long-term soil fertility, these hedgerows can provide the added biomass necessary to do so. In the Sahel, for instance, *Piliostigma*, tephrosia, or more gliricidia, pruned low to the ground, can be grown in hedgerows. Of course, these plants may also require protection from free-grazing animals their first year or two. In areas with more rainfall, a good number of more useful, bushy-type gm/ccs could be used this way.

8. Perennial peanut

The perennial peanut or forage peanut (*Arachis pintoi*) is the cheapest and easiest way to control weeds among perennial crops. It is a ground-hugging, 10-cm thick perennial that looks much like the common peanut, but produces no nuts. It is widely grown in fields of coffee, vanilla, black pepper and oil palm (see photo 43). In urban areas, people often plant the perennial peanut as a lawn, because it never needs mowing or weeding and produces an attractive yellow flower. In several countries, these urban plots have provided planting material for farmers where no other cuttings could be found.

Farmers almost always prefer to use a perennial gm/cc in fields of perennial crops, because they save a lot of labor by not having to replant the gm/cc every year. Perennial peanuts can be planted using seed, but are almost always established using cuttings. This means it requires more work to plant, and some weeding must be done during its first year. However, once the perennial peanut is established and grazing animals are kept out, no more weeding need be done for 30 years or more. Furthermore, the perennial peanut does not climb, so farmers do not have to cut it down around their tree crops.

The one complaint with the perennial peanut is that during periods of drought, it may compete somewhat with young tree crops for moisture, although this problem will automatically be phased out as the soil's OM is built up.

9. Various tree species under the Farmer-Managed Natural Regeneration of trees (FMNR)

Tony Rinaudo, an agronomist and member of Serving in Mission, had been working for years trying to plant trees in Niger. Planting trees was important both to slow down the encroachment of the Sahara Desert and to provide people with a drought-proof income from the sale of firewood and timber. Until 1983, Tony had had no luck at all. Planting seeds or establishing nurseries had failed repeatedly because of the Sahel's harsh environment.

Finally one day, Tony was looking out across the dry, barren landscape, thinking about what he could try next, or whether he should just give up. Suddenly, he noticed a little green tree sprout nearby that was trying to get a start. Then he noticed another one a few feet away. And another.

What Tony discovered that fateful day not only exists all over Niger, but in most of the drought-prone areas of the tropical world. In these areas, there are whole "underground forests" of tree roots, barely visible stumps and dormant seeds (see S94). What those little green sprouts did was remind Tony that these underground forests were still alive; people simply needed to give them a chance to grow.

He tried doing just that. The result: a desert oasis full of trees. And they grew very quickly, because many of the trees already had full-grown root systems.

To make a very long story short, together with World Vision/Australia, where Tony works now, and dozens of other organizations, over 7 million previously barren hectares of Niger have been populated with trees. The same has happened to another estimated 17 million hectares across the Sahel, eastern and southern Africa. That sounds unbelievable, but Tony has satellite photos to prove it.

In fact, some of this work with farmer managed natural regeneration trees (FMNR) has been achieved by smallholder farmers who discovered it independently—a few of them even before Tony did. A lot more of it has resulted from spontaneous spread among farmers themselves from areas where it was introduced by various development programs.[70] But whatever the source, the FMNR revolution is very likely the most astonishing, least expensive success in the history of agricultural development. Even more unbelievable, it has been achieved in one of the most hostile environments for agricultural development on the face of the planet.

All farmers need do to adopt FMNR is to select the trees they most want in their fields, manage them properly and control any free-grazing cattle (see S94).

Large-scale studies of FMNR have now verified that OM levels in the soils have improved, farmers' crop yields have increased by an average of about 50%, incomes have doubled because of the added sale of firewood and women's workloads have been reduced.[71]

Perhaps most important, as a foundational technology, FMNR has provided a platform upon which farmers can develop all sorts of new enterprises. Having soil that is moist, more fertile and protected by the trees' shade from the hot Sahara sun, farmers have begun experimenting with a number of new techniques, new crops and new animals. The biodiversity of the system gives farmers many opportunities to choose how many trees to grow of which species, how to propagate and prune them, and how to use the trees to fertilize and protect the soil and increase their incomes.

For instance, some farmers in Niger and Mali now propagate favored tree species by mixing the trees' seeds with their animal feed and then letting their cattle roam, planting well-fertilized trees all over the landscape. Farmers using FMNR in Ethiopia discovered they can propagate *Faidherbia* trees by cutting the roots. Each root then sends up a new stem from the point at which it is cut—a fact even experienced agroforestry experts had never realized. Some farmers are also grafting improved jujube (*Ziziphus mauritiana*) fruit trees onto the stems of jujube trees that have sprouted from the "underground forest," and are using several species of flowering trees to increase bee-keeping. A top-notch example of such innovation is the Dogon system in Mali (S67). It is basically an FMNR system discovered by local farmers with two additional important innovations: pruning trees in a funnel shape and planting a series of annual gm/ccs among their crops below the trees. Such farmer discoveries add important new benefits to FMNR, and to gm/cc management in general.

The mix of trees that result from FMNR obviously varies, depending on the genetic mix of a region's underground forests and on local farmer preferences. In the Sahel, *Faidherbia*, *Piliostigma*, *Guiera* and *Combretum* tend to dominate, while in other locations, farmers tend to favor other species.

Of course, many trees resulting from FMNR fix no N, and some produce little biomass. Whether or not a particular FMNR system increases soil productivity over the long haul will depend on what percentage of the trees fix N, and how much. The above species favored in the Sahel fix large amounts of N. But farmers make decisions about managing FMNR not just for its impact on soil fertility, but for its diversity of income sources, resistance to drought, and impact on their levels of over-all risk.

The only major downside of FMNR is that farmers have to resign themselves to having at least some lower-value trees in their fields, situated in a more or less random pattern. In the northern Sahel, species propagated any other way may not survive. But farther south where rainfall is more plentiful, farmers tend more often to use animal traction and are able to grow more valuable crops and multipurpose trees. Each of these factors militates against maintaining FMNR in its original form.

Very likely, over time, simple FMNR will come to reign supreme in the driest areas. But as one moves into wetter areas, FMNR will evolve into more intensive, more orderly and more complex systems, using a greater percentage of higher value trees and varied methods of tree propagation and management. In the process, FMNR may well be transformed into something resembling a linear dispersed/intermittent shade system with intensive cropping of diverse crop combinations beneath the trees. Over what latitudes this transition will occur, or what the diverse systems will eventually look like, only time will tell. (This all applies to FMNR only in cropped fields. FMNR can also be extremely valuable in regenerating forests, anywhere.)

But whatever forms it takes, a farmer-managed push-back against the encroaching Sahara Desert is taking shape. It is definitely something to celebrate.

10. *Faidherbia*

Faidherbia (*Faidherbia albida*, previously *Acacia albida*), is a very common tree across the driest parts of Africa. Because it has no leaves during the rainy season, it casts very little shade on crops, creating a dispersed shade with no need for pruning. Smallholder farmers have realized that crops grow better under these trees and leave them in their fields long after they cut down nearly everything else. For this same reason, *Faidherbia* is also the preferred species in some areas where farmers practice FMNR.

Great debates have been waged as to the value of disseminating *Faidherbia* more widely. In areas with less than about 300 mm of annual rainfall, the tree is very useful and should be disseminated. This can be done very easily by transplanting the many seedlings that volunteer or feeding viable seeds to cattle. In areas where the rainfall is higher, I doubt that it would be advisable, because many other species do a better job, from gliricidia to *Piliostigma*.

First, the *Faidherbia* does not fertilize the soil anywhere nearly as well as many other trees. Secondly, the spines of the *Faidherbia* can be dangerous for children, especially in less drought-prone areas where children are not accustomed to constantly avoiding spines. Lastly, the tree grows about one third as fast as gliricidia, so if it is planted from seeds or seedlings, it takes twelve to fifteen years before it really serves as a tree.

Smallholder farmers themselves have already largely "voted with their feet." I have visited a number of sites in Mali, Malawi and Zambia with well over 400 mm of rainfall where plots of full-grown *Faidherbia* had been acting as dispersed shade for over a decade. Yet I have never seen any evidence that the farmer owners have, during all that time, planted any additional *Faidherbia* trees. On the contrary, they have year after year actively killed the new trees that constantly volunteer.

11. Mung beans and rice beans

Mung beans/green gram (*Vigna radiata*, also *Phaseolus radiatus* and *P. aureus*) and rice beans (*V. umbellata*, also *Phaseolus calcarathus* and *Dolichos umbellatus*) belong to the same genus as cowpeas. Like common cowpeas, they do not fertilize the soil very well in unimodal rainfall areas without shade, because of volatilization. Bushy varieties can be intercropped with maize, sorghum or millet, while climbing types should normally only be intercropped with maize. Both species fix about 60 kg of N/ha.

Mung beans usually suffer from fewer problems with insects than common cowpeas, but require a bit more moisture than the cowpea or rice bean. In most of the world, the rice bean is thought to have that name because its seed is the same size and shape of a kernel of rice. In Vietnam, however, people believe it has that name because it is the crop that is grown after paddy rice. It works particularly well in this role, because, compared to the other *Vignas*, it has more resistance to waterlogging and a better ability to thrive on residual moisture.

Mung beans and rice beans are widely grown and have good markets, especially in Asia. People eat them in many traditional dishes, including soups, curries, breads and cakes. Both species are also widely used to make bean sprouts. That most of the N fixed by these gm/ccs is lost because of volatilization in unimodal areas highlights one of the most important reasons that dispersed or intermittent shade systems could be so beneficial in the lowland tropics. A moderate shade could prevent farmers' losing hundreds of millions of dollars of N every year, and would turn these two legumes into valuable gm/ccs.

12. Tephrosias

Both major species of tephrosia (*Tephrosia vogelii* and *T. candida*) are woody-stemmed bushes that reach 3 to 4 m in height under good growing conditions. The tephrosias withstand drought quite well, but sometimes grow poorly below 300 m of elevation. Animals will refuse to eat tephrosia bushes unless they are extremely hungry and the tephrosia is less than a year old,[72] so these species are most appropriate as gm/ccs where animals are allowed to free-graze. Tephrosia plants will die after about four years. Weed control is moderately good if the tephrosia is sown densely.

The tephrosias can also be used to make a powerful insecticide.

13. Peanuts, called "groundnuts" in Africa, and bambara groundnuts

Both peanuts (*Arachis hypogea*) and bambara groundnuts (*Vigna subterranea*, also *Voandzeia subterranea*, *Glycine subterranea* and *Arachis Africana*) are small plants that grow close to the ground. Neither produces enough biomass to do much for the soil by themselves, but they can contribute to biomass and N needs when planted together with other gm/ccs, especially under dispersed shade.

The bambara groundnut is the most drought-resistant of the major edible annual legumes. Farmers generally grow it in areas of frequent droughts.

14. Leucaena

Leucaena (the common name often used for both *Leucaena leucocephala* and *L. diversifolia*) has been called the miracle tree of alkaline soils. It grows quickly and its leaves fertilize the soil every bit as well as do gliricidia leaves. The trees sprout vigorously after being trimmed or cut off, so that a forest of leucaena can last through many harvests of timber.

Nevertheless, the tree presents several problems. First, its ability to drop viable seeds over an area is desirable if one wants a permanent forest, but becomes problematic if one wants to maintain a field. Used as a dispersed shade or contour hedgerow, it can become an invasive weed. Second, termites, rodents or even ants can occasionally kill young leucaena trees. Third, a psyllid native to Central America spread around the world in the 1980s, causing serious problems with the most popular and well-known varieties. One does still frequently see the tree being used as a live fence, but few programs promote it any more. Nevertheless, if something happened to the gliricidia or the moringa, it might well experience a new heyday.

L. leucocephala generally grows better below 800 m in elevation and *L. diversifolia* at higher elevations. Leucaena requires about 600 mm of annual rainfall to grow reasonably well.

In drought-prone areas where the trees already exist, people should be taught to pull open the tender green pods and eat the tender green seeds inside. These seeds cause bad breath and therefore are seldom accepted in normal times, but they could serve as a very valuable famine food. They are high in protein and perfectly edible in reasonable quantities (their mimosine content can cause people's hair to fall out if they are consumed in large quantities).

Leucaena leaves are sometimes fed to animals as a high-protein fodder, but because of their mimosine content, animals should gradually increase their intake of leucaena up to a maximum of about 50%. Only goats will do well on higher percentages of these leaves.

15. *Piliostigma*

Piliostigma (*Piliostigma* spp., previously called *Bauhinia* spp) is a perennial that is sometimes called "cow's hoof" or "ox-hoof" because of its double-pointed leaf. It grows wild in lowland, semi-arid to sub-humid Africa, including the entire Sahel and most of eastern and southern Africa. It can also be found in at least a dozen countries of Latin America and Asia. *Piliostigma* withstands drought and pressure from grazing animals so well that it often constitutes as much as 20% of the natural biomass in the central to northern Sahel. It normally grows as a large bush, but can also grow naturally or be pruned to become a medium-sized tree.

The *Piliostigma*'s leaves may well increase subsequent crop yields more than those of any other legume native to Africa. I once did a participatory rural appraisal (PRA) of local legumes in about 40 villages in Burkina Faso and Mali. In each case, I asked a group of women to gather branches from the various plants that made their soil more fertile and then arrange them in a line on the ground, in accordance with their impact on crop yields. Out of some 20 to 25 plants, the women in every single village placed the *Piliostigma* in the number one position.

Because they recognize *Piliostigma* as a particularly valuable gm/cc, farmers in the Sahel traditionally use it in several interesting ways. In parts of Burkina Faso, farmers harvest the leaves from out in the bush and hand-carry them to their fields. In northern Ghana, when the fallow period ends, they pile soil over the *Piliostigma* plants right before burning the fields, to protect the *Piliostigma* leaves from being burned. In parts of Burkina Faso and Mali, farmers selectively weed, leaving all the *Piliostigma* bushes that volunteer in their maize and millet fields. Then each year right before the rains start, they prune the bushes to ground level, spreading the prunings over the soil. Women often pile the larger branches on nearby termite mounds so the termites will cut the leaves off the branches. Then the men apply the leaves to the soil and the women take the branches home for firewood.

A number of other characteristics make *Piliostigma* a remarkable resource. Being a perennial, it can last for decades in a field. The *Piliostigma* may succumb to grazing during its first dry season if the animals are desperate for fodder (usually only in areas of 400 mm of rainfall or less), but after that, it is completely resistant to animals—this being crucial to survival in the Sahel. Each year, two or three seedlings customarily volunteer near each *Piliostigma* plant. This spontaneous propagation is not enough to make *Piliostigma* a weed, but it does allow patient farmers to regularly transplant new seedlings each year to an alley cropping hedgerow or another desirable site, without making a nursery.

Photo 20. Mali. This *Piliostigma* has been pruned as a tree. *Piliostigma's* ability to fertilize the soil in extremely drought-prone areas is unmatched by any other native African plant.

Probably the biggest problem with *Piliostigma* is that, if grown from a seed or a seedling, it grows very slowly during its first three or four years. Presumably it does this to grow a large root structure.

Richard Dick, a researcher in Senegal, has found some truly amazing properties of this plant. During droughts, the *Piliostigma* draws water from deep in the soil and exudes it near the surface so the water can dissolve more nutrients before being reabsorbed by the plant. Nearby crops can take advantage of some of this pumped water loaded with nutrients.[73]

We are still looking for the best ways to use *Piliostigma* as a gm/cc. I suspect that it will work very well in a dispersed shade system as a hedgerow under every second or third row of gliricidia trees (leaving at least 15 m between hedgerows). It would add enough extra biomass to the system to increase yields over time. Additionally, if a gliricidia tree died, a nearby *Piliostigma* bush could simply be pruned as a tree to take the place of the dead one.

16. *Sesbania rostrata*

Sesbania rostrata is widely used as a traditional green manure in rice paddies. As such, farmers plant it at the first rains, it grows very fast and fixes N quickly, and then the farmers cut it down at the flowering stage in order to plant the rice. The seeds, flowers and leaves of the *Sesbania* are all edible, although the seeds must be soaked before cooking. However, as land becomes scarcer and rains more irregular, farmers will be less and less willing to use traditional green manure systems because they reduce the time in the fields when their primary crops could be growing.

17. Soybeans

Soybeans (*Glycine max*) are widely grown in China, Vietnam, and other parts of Southeast Asia, because of their popularity as a traditional food. They are usually mono-cropped or grown as an intercrop with maize.

Outside of these areas, programs have attempted to introduce soybeans among smallholder farmers because of the beans' protein content of about 40%, but often the crop has not been very popular. Even when smallholder farmers understand soybeans' food value, what interests farmers is not so much the percentage of protein in the seed, but rather the cost of each 100 gm of protein. Soybeans normally require weeding, and often fertilization. And the bean does not taste good merely boiled, so it must usually be prepared as tofu, soy milk, or some other labor-intensive product. As a result of all the labor required to produce and prepare soybeans, each 100 gm of soybean protein costs a lot more than that of other gm/ccs. Besides, just as with other short-lived annuals, soybeans are not a very good gm/cc because they suffer from volatilization, poor synchronization and low biomass production.

In some places, though, all this is changing. In Malawi, a soybean porridge that is very simple to make and quite tasty has become popular across much of the country. Also, as animal feed factories slowly proliferate across the landscape, soybeans command a good price in more and more markets, making them more profitable as a gm/cc. Furthermore, some new, taller varieties produce more biomass, while the "promiscuous soybean" (given this name because it can use a wide variety of rhizobia) reduces the species' need for fertilizer. Soybeans have the potential of becoming a much more important gm/cc in the future.

The Minor Lowland Species

1. Lima beans

Lima beans (*Phaseolus lunatus*) are grown in much of Latin America and Africa. Farmers grow them as an intercrop with maize, as well as in home vegetable gardens and on garden fences. Both bushy and viny types exist. The viny types tend to prefer humid conditions, where they sometimes serve as the principal grain legume. The seeds, tender pods and leaves are edible, but varieties with darker-colored seeds tend to contain toxic levels of hydrocyanic acid. Before consumption, these latter varieties must be boiled, with several changes of water. With more good research, lima beans might be an attractive gm/cc species, especially where they have a good market.

2. Crotalarias

Crotalarias (*Crotalaria* spp.) are used in Brazil by hundreds of thousands of medium- to large-scale farmers. Tanzania is the only place I know where they are used widely by smallholder farmers. There, farmers intercrop sunn hemp (*C. juncea*) with maize, both to fertilize the soil and to provide an insecticide used for stored grain. Once grown by tens of thousands of farmers, its use decreased significantly when the subsidies supporting its use ended and fertilizer companies supported a campaign to get rid of it.

3. Tithonia

Tithonia (*Tithonia diversifolia*), or wild sunflower, is not a legume, nor does it fix N. However, it is a phosphorus accumulator. It accesses soil P that is not available to other plants, making that P available to crops when its leaves decompose. It can therefore be used profitably in fields where the lowest maize leaves are purple on the outer edges.

Tithonia plants can be found along many highways in Latin America and Africa at elevations of between about 400 and 800 m, especially when their bright yellow, daisy-like flowers are blossoming. Farmers can easily collect seeds from these wild plants and plant them in fields as contour hedgerows (S45) or broadcast them as a relay crop (S42).

In the Philippines, tithonia leaves are traditionally gathered and mixed with animal manure for fertilizing paddy rice fields at transplanting time. The plant is also used there to eliminate imperata grass (S12).

4. Yam beans (*Pachyrhizus erosus*), Stylosanthes (*Stylosanthes* spp), and some Desmodiums (*Desmodium* spp).

Each of these species is used as a gm/cc in a few places, but none provide enough benefits to be of wider use. Yam bean tubers enjoy a good market in Southeast Asia, but lacking that market, would probably not be accepted elsewhere. Stylosanthes has been tried many places, but with limited adoption, because it produces no human food or other benefit. Greenleaf desmodium (*Desmodium intortum*) is being promoted in the "push-pull system" for fighting striga and stem borers in East Africa. It may become much more widely used if it is found to effectively control the fall armyworm, which recently invaded Africa for the first time.

Highland Species

Most lowland gm/cc species start losing their vigor—and therefore their usefulness as gm/ccs—at around 1,500 m in elevation. Farmers have used a few of them up to 1,800 m, such as the lablab beans grown near Siguatepeque, Honduras, but these farmers may not have known how well runner beans or fava beans grow at that altitude.

Luckily, a half-dozen highland gm/cc species can do a very good job from 1,500 m on up to 3,000 or 3,500 m in elevation. Tarwi grows well up to over 4,000 m. The following species are the most important highland gm/ccs, roughly in order of descending importance.

1. Runner beans

Runner beans (*Phaseolus coccineus* also *P. multiflorus* and *P. polyanthus*), also called scarlet runner beans, Seneca beans and butter beans, are probably the most popular highland gm/cc in the world (photo 21). Before the Spanish arrived in the Americas, these beans were grown throughout the two continents' cooler environments, from central Chile along the Andes north through Colombia, then picking up again in Nicaragua through Mexico and from upstate New York into Canada. In Africa and Asia, I have found them in highland areas of Ethiopia, Kenya, Uganda, Rwanda, the Democratic Republic of Congo (DRC) and Nepal.

The runner bean generally produces even more biomass than does mucuna. Farmers in central Honduras have increased their soil's fertility for 20 years straight by intercropping runner beans with maize without using any synthetic fertilizer.

Runner beans sometimes grow poorly the first year in a given field because of a lack of appropriate rhizobia. After that, they usually grow quite well because the presence of plants' roots in the soil increases the population of the rhizobia. Runner bean plants produce a tuber that sprouts each year for up to ten years, thereby eliminating the cost of reseeding, but farmers tend to replant them about every fifth year. In Latin America, runner beans almost always enjoy a higher market price than common beans (*P. vulgaris*) because people prefer their taste, although the reverse is true in most of East Africa. The pods, leaves and flowers of the runner bean plant are edible.

Photo 21. Guatemala. As this photo shows, runner beans can outdo even mucuna in their production of biomass/ha. In contrast to mucuna, however, these beans are edible. The seeds are larger than most beans, and are either white or a variegated purple and black.

When intercropped with maize, the excessive weight of the runner beans often pulls down the maize. In some cases, farmers have to limit the runner bean to 1 seed/5 sq m of land. In fields below 2,000 m in elevation, tephrosia or pigeon pea might be intercropped in the same field to help support the runner bean's weight, so that it could be planted more densely.

2. Fava beans

Fava beans (*Vicia faba*), also called broad beans, are an excellent gm/cc. It can fix up to an astounding 400 kg/ha/year of N. Fava beans are grown in North America, Central America, the Andean countries, Europe, East Africa, the Himalayas and parts of China and Southeast Asia. The plant is quite drought-resistant, but does not control weeds very well. The beans, which are very tasty, can be eaten green (boiled like lima beans, or in soups), fried (usually as a snack food like peanuts), or roasted (creating a jaw-breaker-like snack food).

Most people do not realize that lowland varieties of fava beans also exist. The Nubian people, who live at 300 m in elevation along the edge of the Nile River near the Egyptian-Sudanese border, grow them and consume them as their main staple. These lowland fava beans are, I believe, completely unknown to the rest of East Africa, not to mention Latin America and Asia. The introduction of these lowland varieties to these other areas could provide tremendous benefits to people who already appreciate the bean but cannot grow it.

Over-all, fava beans are a widely known, highly appreciated species. They can be used in a multitude of ways: as an intercrop with almost any plant more than 1 meter tall, as an understory for tree crops, as an edible and marketable monocrop in rotation with other crops, as a hedgerow species in alley cropping systems, etc. Given the relatively small number of edible gm/cc species that are adapted to high altitudes, such a versatile species is exactly what the tropical world's highlands need.

3. Lupines

Among the lupines, both tarwi (*Lupinus mutabilis*) and tailcup lupine (*L. caudatus*) are used as gm/ccs. Like fava beans, tarwi can fix up to 400 kg/ha/year of N—the equivalent to the N in over 19 sacks of urea weighing 45 kg each! In fact, when farmers intercrop tarwi with other crops, they must take care to keep the plant population of the tarwi below 2,000 plants/ha. Otherwise, the excess N will cause maize and other crops to grow an excess of vegetative mass instead of grain or fruit.

Tarwi is an erect plant that reaches about 1.5 m in height and withstands the very cold conditions of the high Andes, anywhere from 2,500 to 4,000 m in elevation. The grain is widely eaten in the Andes but must be left in running water for a couple of days before cooking, to rid it of anti-nutritional factors.

Tailcup lupine, similar in many ways to tarwi, is grown across the Ethiopian highlands. As with tarwi, the grain of this species is edible after being soaked in running water.

4. Alder trees

Alder trees (*Alnus acuminata* and *A. nepalensis*) are non-leguminous trees, but still fix N through the activity of soil bacteria called actinomycetes. Asian farmers use *A. nepalensis* in dispersed tree systems (see S117), while in Mexico and the Andes, *A. acuminata* is increasingly used in reforestation projects. Both species grow quickly at high altitudes and are resistant to droughts and poor soils. They are used as fodder for sheep and goats, as well as being effective gm/ccs for soil fertility maintenance.[74]

Alders, like gliricidia, re-sprout easily after being pollarded, so they work very well as dispersed trees.[75] At the altitudes that alders grow, their shade alone cannot improve the production of field crops, so their use is motivated by the other advantages of dispersed trees: their contribution to maintaining soil fertility, protection against drought, and the value of the timber, firewood and fodder, all achieved without a reduction in crop production.

5. Chickpeas

Chickpeas (*Cicer arietinum*), also known as garbanzo beans, are grown widely in Ethiopia in dry season rotations with other crops (S112), and are grown and consumed around the world, especially in India, the Middle East and the Mediterranean. This kind of edible legume that can grow on residual moisture has tremendous potential to fit well into numerous highland farming systems around the world.

6. Sweet clover

Sweet clover (*Melilotus albus*), also called honey clover and white sweet clover, is a very drought-resistant plant with both annual and perennial varieties. It is so hardy that we refused to spread the S103 system for about 10 years until we learned of a way to get rid of it without using herbicides. Eventually we learned that it would die if cut off at ground level at the *beginning* of a 6-month dry season.

7. Minor highland gm/cc species include vetch (*Vicia* spp.), grasspeas (*Lathyrus sativa*), buckwheat (*Fagopyrum esculentum*) and common peas (*Pisum sativa*).

Vetches are more popular with large-scale farmers than with smallholders, largely because they have no real advantages to offset the fact that they produce no human food. The one place I have seen a vetch used widely among smallholder farmers is in Michoacán State in Mexico, where farmers traditionally intercrop *Vicia toluca* with their maize, and then feed it to cattle throughout the dry season (S105).

Grasspeas are used as a traditional green manure or are rotated with teff (S114) in highland Ethiopia, but other gm/ccs provide both better food and better fertilizer. For instance, runner beans can be intercropped with maize one season, and the teff grown the next. The grasspea is considered more drought-resistant than lupines (*Lupinus* spp.), but less drought-resistant than chickpeas. Generally, the grasspea produces more biomass than either the chickpea or the lupines, and this biomass can serve as fodder, as well as a gm/cc.

Grasspeas are eaten in Ethiopia, even though the traditional varieties contain a dangerous toxin. For farmers who want to eat the grain, the International Center for Agricultural Research in the Dry Areas (ICARDA) has developed low-toxin varieties that are safe for human consumption.

Buckwheat is grown in Rwanda and Brazil, but it produces relatively little biomass and is sensitive to frosts. Common peas are used in southwestern Uganda as soil conservation barriers, with the added advantage of being edible and working as a gm/cc. But dwarf Napier grass holds the soil much better than peas,[76] and other gm/ccs could improve the soil more.

Species to Avoid

Two categories of gm/ccs should never be used except where they already exist, because of the problems they cause. The first includes thorny or bothersome species. You will understand why they should be avoided if you have ever seen farmers trying to bury dangerous thorns in their fields, or you have seen a child's foot that has become badly infected after being punctured by a thorn. Thorny species, including many *Acacias* and the *Faidherbia albida*, that commonly drop their thorns on the soil, can be used largely without problems in drought-prone areas where sharp spines are common and children are used to avoiding them, but should not be used in areas where thorns are uncommon.

The itchy mucuna is another very bothersome plant to be avoided. It is true that thousands of women in places like Mozambique and southern Malawi actually go into forests to collect the grain of the itchy mucuna to be eaten. As with regular mucuna, they have to boil these seeds six times for several hours each time to avoid the negative effects of the L-Dopa. In such cases, it is very urgent to show them how to grow some other legume so they do not feel such a need for protein that they must do this painful work. Similarly, farmers presently using the itchy mucuna as a gm/cc should immediately be sold some non-itchy mucuna seed or told where they can obtain it.

The second category of gm/ccs to avoid includes invasive species that, when introduced into a new area, can seriously damage the local environment or human livelihoods. We need and appreciate our gm/ccs' abilities to grow with tremendous vigor, be propagated easily and require little human care or attention. Frankly, we want them to grow like weeds. But some species overdo it. They not only grow like weeds, they *become* weeds—of the worst kind. Eventually, they may infest huge areas, making food production more difficult or even creating wastelands out of previous farmland and pastures.

Among these problematic species are the two kudzus: tropical kudzu (*Pueraria phaseoloides*) and common kudzu (*Pueraria montana*). Common kudzu was imported into the United States and has spread across large swathes of the nation's Southeast, causing serious ecological and economic problems. Its cousin, tropical kudzu, has caused serious damage across parts of Latin America. In these areas, it has become the weed that people most love to hate.

Other invasive species that have caused serious damage in one environment or another include the perennial soybean (*Neonotonia wightii* or *Glycine wightii*), lantana (*Lantana camara*) and calopo (*Calopogonium mucunoides*) (see S2).

Some of these species have caused problems in one area but not another. Knowing in which areas a plant will be problematic is not easy. Climate and rainfall patterns are not the only indicators. For instance, lantana is considered a terrible pest by large landholders in much of South Africa, but in Zimbabwe, smallholder farmers gladly plant it to make live fences. The difference in this case seems to be that in Zimbabwe the lantana serves as easily managed, permanent live fencing, needed more intensively because landholdings are small. Also, the country's higher population density provides plenty of man-power to keep the lantana under control. Consequently, the benefits outweigh the added labor cost.

S2. Any crops/calopo (*Calopogonium mucunoides*). NEVER. Ghana. A president of Ghana introduced calopo as a cover crop, but it soon spread over large parts of Ghana and became a very invasive weed. Many farmers now bitterly ridicule him by calling him "President Calopo."

Some people might feel that some of the other species could also provide major benefits. However, in the cases of both the thorny and invasive species, other gm/cc species can bring all the same benefits without becoming a pest. For instance, the perennial peanut can accomplish everything the kudzus can, and do it better. First, there is no danger that the perennial peanut will become invasive. Second, it does not climb trees, so farmers do not have to keep clearing it away from tree trunks. Third, perennial peanuts form a cover that is only about 10 cm thick, so that if fruit, or anything else, falls to the ground, it is easily found. Fourth, snakes cannot hide in its thin groundcover, as they can with kudzu.

The danger that these invasive and thorny species will cause problems is just not worth the inferior benefits they might provide. **None of these species should ever be introduced into new areas.**

The Most Popular Lowland Species

For the last six years, in eight African countries, I have been doing a study to see which of all these scores of gm/cc species have become the most popular among smallholder farmers. The criterion I used was how far and how fast the farmers themselves have spread these species to other farmers and other villages. The results were a surprise to me. There were four species that were, by a huge margin, the most popular. These were pigeon peas, lablab beans, jack beans and gliricidia. In most cases, future programs in lowland maize-growing areas should start out working with these species, or if already active, seeing these species as the ones that should most likely be added to the species they are already working with. I suspect that Lojy Be and crawling cowpeas will eventually occupy two more spots on this list, but they are not widely spread enough at present to be able to tell.

Chapter 7. Achieving Adoption of Gm/ccs

A Debate on its Deathbed

A great deal of discussion has centered around the levels of adoption or abandonment of gm/cc systems that have occurred around the world. Several traditional gm/cc systems are slowly being abandoned because of a number of factors. Conventional chemical technology is spreading, chemical inputs are becoming more widely available, traditional foods have become unfashionable, some professional extension workers criticize farmers for doing things that are discouraged by fertilizer companies or that are not recommended in university textbooks, maize fields are being converted to other uses, and large-scale farmers or land-grabbers buy or steal smallholders' lands. This gradual abandonment has affected such previously widespread systems as the 1st maize/rice bean (S35) and 1st maize/cowpea (S23) systems that once existed from Mexico through Nicaragua, and the 1st maize/runner bean system (S97) that once stretched most of the way from southeastern Canada to Chile.

At the same time, other gm/cc systems continue to expand widely and quickly, right up to the present day. One is the 3rd maize/mucuna system (S52) that has spread spontaneously through parts of Mexico, Guatemala, Belize and Honduras over the last 60 years. However, while it continues to spread in several parts of Mexico and the El Peten Department in Guatemala, it is now receding in Honduras and has nearly been lost along Guatemala's Polochic River.[77] As a result, the debate has persisted for decades: will gm/ccs become sustainable, or will they, after much blood, sweat and tears, become just another footnote in history?

In 2009, the development organization World Renew asked me to do a six-nation study of agricultural work in Africa. One day in September, 2009,

I found myself sitting next to the village chief and his elders in Koboko village, Malawi. In front of us, 30 women and their children were gathering under a huge shade tree, the traditional site for village meetings. Having been told that a strange white man wanted to ask them some questions, they were gradually squeezing together on an assortment of hand-woven mats and tiny rough-hewn chairs.

I began my little interrogation with the one question I ask more often than any other: "What is the most important single problem that prevents you from having enough food to feed your children?" Without even waiting for one of the village's male authorities to answer, one of the taller women spoke up: "Our soil is tired out. And it is getting worse every year."

Before she had finished speaking, four or five other women chimed in, "Yes, what she said is true!"

"Last year I harvested 35 bags of maize. This year I only harvested 27, even though it rained a lot."

"We no longer have any way to keep our fields fertile."

"Our soil has become so hard that even when it rains, the water just runs off."

When things had quieted down, the village chief, calmly and authoritatively, put his official stamp of approval on the obvious consensus by voicing his heart-felt agreement.

Frankly, I was totally taken aback. Just five years earlier, Malawi had suffered one of Africa's worst droughts ever. People were so hungry they were cooking up and eating the bark from trees. Many died, and millions would have if food aid had not been distributed throughout the country. Yet in this village, everyone agreed that soil fertility was a worse problem than droughts. They explained that, sure, the drought had been horrible, but droughts only occurred once every five or ten years. The lack of soil fertility threatened to destroy their food supply—permanently.

The women were unanimous. They were adamant. And they were obviously scared. Even though they were desperately poor, they had never before faced such a long-term and apparently insoluble threat to their very survival.

Everywhere I went in lowland, drought-prone Africa, the answer to my question was the same. Harvests were decreasing from 5 to 25% every year. But why were they facing such a crisis, and why for the first time?

The fact was, they had been hit by a perfect storm of problems. First, they no longer had enough land to fallow it. Fallowing had been the technology African farmers used for 3,000 years to maintain their soil fertility. But population growth had reduced their farm size to less than 2 ha. You cannot fallow ¾ of a 2-hectare farm and still grow enough food for your family to survive. So fallowing across sub-Saharan Africa was in its death throes. Second, farmers could not use fertilizer to maintain their soil's productivity because wealthier countries had used up the world's cheap energy, and the manufacture of nitrogenous fertilizer requires a tremendous amount of natural gas. Synthetic fertilizer was now too expensive to be profitable. Third, global warming was making rains more and more irregular.

Every one of these problems was hitting farmers for the first time in history. The crisis was serious.

The same crisis has affected the farmers in drought-prone lowlands across the developing world. In East Asia, with its thriving economies, many people just left their farms to find good jobs in the cities. In Latin America, people left their farms, but wound up with low-paying jobs in semi-permanent slums that are swelling to overflowing. In Africa, swarms of youth have moved to the slums, but jobs are scarce to non-existent; the people left in the lowland, drought-prone areas are being hit hard.

I realized that a major famine was in the making, and I had to do something about it. I wrote a chapter for the book, *State of the World 2011*, called "Africa's Soil Fertility Crisis and the Coming Famine." In that chapter I predicted that "within the next four or five years," a famine would begin that would affect tens of millions of people in the lowland, drought-prone areas of the continent.[78] The famine has now officially started—in early 2017. In fact, upon announcing the famine, the UN called it "the largest humanitarian crisis" since 1945.[79]

My prediction was off by less than four months, and the affected area coincided exactly with the area I had predicted, except for the highland areas of Ethiopia. I did not make that prediction based on some extensive knowledge of weather patterns or global warming; I made it based on a knowledge of soil conditions. A lack of rainfall is not what is causing the increasingly frequent "droughts" affecting Africa (although the rains *are* becoming more irregular because of climate change). These droughts are caused by lack of SOM and a resulting lack of rainwater infiltration that occurs when soils are bereft of OM. Rainwater infiltration has fallen from about 60% to 20% in many areas. I have gone out into many fields in the Sahel after 3- to 4-hour downpours and dug down into the soil. It is usually moist down to only about 5 cm in depth. That near the surface, the tropical sun dries it out completely in two or three days.

What makes this drought unique, and particularly dangerous, is that true droughts, caused by a lack of rainfall, are naturally self-terminating. Once it starts raining again, the drought is over and the relief efforts can move on to some other emergency. But with this drought, nothing will terminate it except a change in farmers' practices. This drought will continue to worsen (with some annual variations) until farmers start increasing the SOM in their soils. Africa's Great Famine will continue to get worse until farmers, *en masse*, start using gm/ccs.

Fortunately, the world is already awakening to Africa's soil fertility crisis. The farmers themselves, of course, are most aware of the problem. For years, up through most of the 20th century, a large majority of African farmers really were not very interested in gm/ccs. In the early 1980s when I would tell African farmers about gm/ccs, they would usually respond with something like, "By the way, we have a problem with an insect pest. Can you help us with that?" Now, when I describe gm/ccs, farmers' first response is, "When will you bring us some seed?"

Today, most smallholder farmers in lowland, drought-prone Africa, like the women in Koboko , know something is terribly wrong. Their food supply is diminishing, and few of them know of any way to solve the problem. Scores of NGOs are aware of the same phenomenon and its cause.

Recently, a fair number of articles on the subject have been published, but only a very few of them have related the soil fertility crisis to the "drought."

Nevertheless, there is more evidence than a risky but surprisingly accurate prediction to make the connection between soils and droughts. The most important evidence is that in countries such as Honduras, Mali, Zambia, Malawi and Madagascar, thousands of fields in which farmers have built up the soil's OM by using gm/ccs show no signs of suffering from the "droughts."

As the soil crisis deepens, fewer and fewer field people in Africa believe that gm/ccs will ever become just a footnote in history.

Characteristics of Successful Gm/cc Systems

If we are going to make gm/cc systems as sustainable and widely popular as possible, we need to understand what makes certain gm/cc systems so much more popular and so much more widely adopted than others. Most of the programs and organizations that have been successful in introducing sustainable gm/cc systems have followed a small list of guidelines:[80]

1. **The land to be occupied by gm/ccs that produce no food or income, must have no locally known alternative opportunities for producing food or cash crops.**

This means that if the gm/cc system does not produce food or an income, it must occupy land that cannot be used for producing either food or an income during the time the gm/cc would be using it. That is, it must be produced on land that has no known significant opportunity cost. This sounds like an almost impossible rule to abide by, but we are finding more and more ways that gm/cc systems can live by it:

a) If the gm/cc species does produce food or an income, it may be grown in any way that fits into the system, like any other equally valuable food crop. Nevertheless, the changes to the previous system should be held to an absolute minimum.

b) If a gm/cc species does not produce food or an income, it can be intercropped with or relayed into an existing crop that is not already intercropped. For instance, farmers intercrop bushy-type jack beans in maize, sorghum, millet, cassava and coffee fields. Intercropping has always been the most popular way of growing tropical gm/ccs.

c) Farmers can grow a gm/cc on a wasteland to restore it, or during the first year or two of a natural fallow. Farmers in Cameroon and Vietnam, for instance, broadcast tephrosia seeds into their first-year fallow (S8). By doing this, they cut the fallow period by three years. As a result, they increase by 75% the amount of land they can plant each year to food crops.

d) The gm/cc can be grown mostly during the dry season, in any of three ways. First, farmers can relay the gm/cc into traditional crops. Examples of this are the 2nd maize/cowpea (S24) and 2nd maize/lablab (S16) systems in Thailand. Second, farmers can plant the gm/cc after the harvest of normal crops, as in the case of the rice/rice bean system in Vietnam. (S37) Third, they can intercrop the gm/cc with the normal crop and then let it grow through the dry season after the traditional crop is harvested, such as with the maize/sweet clover system in Mexico (S103).

e) The gm/cc can be grown as a dispersed or intermittent shade *over* virtually any other lowland tropical crop. Both the Dogon intermittent shade system (S67) and the gliricidia-based dispersed shade system (S3) stand as examples of this approach.

f) Farmers can grow a gm/cc under fruit or forest trees, or under almost any perennial crops.

g) At the beginning of the growing season, most crops are very small and therefore do not occupy all the space around them. Short-term gm/ccs like the 60-day cowpea can use these spaces to grow.

h) Occasionally farmers can use other ways of growing gm/ccs without affecting the main crop. These include growing gm/ccs during periods of probable frost, as farmers do with lupines like tarwi, or growing gm/ccs in soils that are too acidic for their crops, as they do with mucuna or buckwheat (*Fagopyrum esculentum*).

2. **Gm/cc systems that produce no food or income must cost very little or no money.**

This rule implies three things. First, farmers must be able to produce their own gm/cc seed year after year. Second, the gm/cc cannot require the use of a commercial inoculant (a product that increases N-fixation). Third, the gm/cc must have no disease or insect problems that significantly diminish its growth and cannot be controlled by inexpensive natural means. If a gm/cc species cannot fulfill all these three conditions, we must find another species.

3. **Gm/cc species that produce no food or income must not increase net labor costs very much.**

This rule means that gm/ccs must usually be left on the soil surface. It also makes intercropping gm/ccs particularly advantageous; the gm/cc often provides weed control that saves more labor than that needed for managing the gm/cc.

Reduction of labor is a major reason why farmers appreciate zero till systems. Since zero tillage works best when farmers maintain high levels of SOM, this desire to eliminate tillage is an important motivation for farmers to use gm/ccs.

4. **Gm/ccs that produce no food or income must fit into the existing farming systems.**

Farmers will almost always see gm/ccs that do not produce food or an income as being much less important than crops that do. Therefore, farmers and extension workers must generally adjust the gm/cc to fit into the existing cropping system, rather than change the farming system to fit the gm/cc.

5. **The gm/cc system should provide at least one important benefit in addition to improving the soil.**

In a worldwide study COSECHA did of introduced gm/cc systems, virtually every system that persisted at least five years after the development organization left the area produced important benefits above and beyond soil improvement. We should always try to promote gm/cc species that people can eat, that animals will eat or that produce some other highly valued benefit.

In conclusion, good gm/cc species should establish themselves easily and grow rapidly under local conditions. They should be able to fix plenty of N. They should also be resistant to insects, diseases, grazing animals, droughts, or any other problem they may have to face, and they should produce enough seed for future plantings. If the gm/ccs are to be used for intercropping, they should withstand shade and fit in with the growing cycle of the main crops. And they should provide major benefits beyond just fertilizing the soil and improving its moisture retention. Virtually all the gm/cc systems recommended in this book fulfill these guidelines.

A Useful Message

In those areas where fallowing ended within the last couple of generations (that is, in most of Latin America and Africa), a great way to introduce gm/ccs is to start by asking farmers, "What did your parents and grandparents do when the soil in a plot of land became weak or infertile?"

After some thought, and perhaps a few false starts, they almost always say something like, "They left that land and went to find another piece of land that was fertile." You can then carry on the conversation something like this:

"So what happened to the land they left behind?"

"The forest (or grass) grew back."

"Did your parents/grandparents ever come back and use the land they had abandoned?"

"Yes, after ten or twenty years."

"The soil there was fertile again?"

"Yes."

"So soil can become fertile again after it has become weak?"

"Umm, yes, I guess so."

"And did your grandparents use fertilizer to make it fertile again?"

"No! Our grandparents never even knew what fertilizer was!"

"Then what happened? How did that soil become fertile again?"

At this point there is usually some discussion, but before long, people will agree that the soil became fertile because lots of leaves dropped on it and rotted, and those rotted leaves helped form new topsoil.

"And why don't you do the same thing now?"

"Because we don't have enough land. If we leave ¾ of our land to grow a forest, we'll never have enough to eat!"

"OK, so **what we are going to learn now is how to grow the plants from the forest that are best at making our soil fertile, but in the same fields where you are growing your crops!** You can grow these plants right in between your crops. In the end, we will produce just as much or more food as you do now, but you will be fertilizing your soil at the same time."

At this point, farmers usually start asking for some seed.

Finally, we can mention the other advantages that these "seeds from the forest" can provide when grown in their fields. We should emphasize that this technology is nearly the same one their grandparents used. Their grandparents were not ignorant! We have not invented something totally new; we are just making a small modification to what their grandparents did—a modification that has only become necessary in the last 30 years.

The Smallholder Farmers' Role

Many people working in agricultural development strongly believe that farmers should participate in their own agricultural development. Actually, we need to go one step further—farmers should become the protagonists of their own development. As Paulo Freire, the great Brazilian prophet of *concientización*, famously wrote, they should become "the authors of their own destiny." A healthy program goal is that after a good technology has been found and the dissemination process started, farmers should be leading the process within five or six years.

Photo 22. Guatemala. The best trainers of smallholder farmers are other smallholder farmers who have already successfully adopted the technologies involved. The simple instrument demonstrated here is used to mark the position of the crop rows that fall between the lines drawn with an "A" frame level.

Achieving authentic farmer participation is not as easy as it sounds. Is everyone participating, or is it mostly the powerful? How can programs make sure the poorer people are not being manipulated by the powerful? How can programs prevent villagers from just voicing answers they think outsiders want to hear, or that they believe are the "correct" answers? How do programs keep men from doing all the talking, or go against cultural norms—for example, the belief that a woman should never contradict anything a man has said? Can we assume that farmers know all the possible technological solutions to a given problem? If not, can they effectively participate? In most cases, we need to keep working for years at improving the quality of people's participation.

Most of the technical questions that arise in this book should be discussed with the farmers and most of the answers should be provided by the farmers (to the extent they know about the solutions), in an atmosphere of complete mutual trust and confidence.

We should never assume that the gm/cc cropping systems recommended here are the only possibilities in a given situation. Over time, program leaders, extensionists, or smallholder farmers may find better gm/cc species or better ways of using them. Without doubt, dozens of good gm/cc species and hundreds of good gm/cc systems, have yet to be discovered. The world definitely needs more gm/cc systems. Farmers and programs should always be looking for new gm/cc species and new ways of using them, just as farmers should always be looking for more crops to grow and better ways to grow them.

Do not ever allow this book's recommendations to reduce people's creativity. We must encourage creativity to blossom. But neither should we underestimate the difficulties involved in finding successful gm/cc systems. Scores of programs have tried and failed.

Photo 23. Honduras. These smallholder farmer leaders are listening intently to another farmer leader as he explains a new technology. They are vitally interested because they saw the technology in practice during an educational field trip, and they feel confident it could also make a major difference in their own lives.

Some people may wonder, "Why should we become involved in finding solutions to the smallholder farmers' problems? Don't they have huge amounts of indigenous technical knowledge?" Indeed, they do. But if they already know the best solutions to their problems—solutions that are within their means—why have they not solved those problems already?

No one can argue that they are not smart enough to do so. After all, smallholder farmers have spent hundreds of years learning about technologies and cropping systems they needed at different times. However, during all those centuries, they could use fallowing, so there was little need to learn anything more about maintaining soil fertility. Only in the last 20 to 50 years has population growth brought an end to fallowing, and much of that time, synthetic fertilizer was quite cheap. Only in the last 10 to 30 years have farmers really needed to learn about something like gm/ccs, and 10 to 30 years is not enough time for smallholder farmers to learn, through informal experimentation, everything they need to know about such a complex subject. Besides, none of those farmers had access to the seeds of more than just a few of the gm/cc species in use today.

What, then, should be our role as people who wish to promote true farmer-protagonist, farmer-led development? Certainly, we should be well-informed, we should listen carefully to the farmers, we should value their experience and we should be very slow to ever discount their knowledge and priorities. We also need to learn how the technologies that smallholder farmers have selected can best be adapted to local needs and how to most effectively communicate or promote them.

Then we need to train farmers in the relevant gm/cc systems mentioned below, while respecting their priorities, their knowledge and their abilities.

The training process should become a dialogue in which farmers and outsiders both provide information and learn from each other. Eventually, by experimenting on their own, by learning to train other farmers, and by having access to information like that in this book, farmer leaders should take over the development process themselves. In fact, they *must* do so, if the process of agricultural development is ever to become locally sustained.

Finally, one way to start farmers experimenting is to only give approximate information on crop spacing and planting. Let farmers know they should experiment with various possibilities on either side of the recommended averages. Farmers will thereby become more involved in the process of innovation right from the very start, and will become more convinced of the value of their own experimentation. After all, we can never tell farmers the exact spacing and planting dates that would be best for a given intercrop. The ideal spacing and dates for each farm will be influenced by crop varieties, ambient temperatures, soil fertility, amount of rainfall, weed populations, availability of labor and farmer priorities. For these reasons, this book either does not give these data or gives only approximations.

My book, *Two Ears of Corn: A Guide to People-Centered Agricultural Improvement*,[81] explores these issues of extension methodology in much more depth.

Collecting the Necessary Information

Through farmer interviews

The first step in choosing one or more gm/cc systems for a particular area is to get to know the area by talking with farmers. Whenever possible, ask open-ended questions. Allow farmers to digress, bring up issues they want to mention, ask questions and express their opinions.

The essential information includes the following:

1. Do the farmers have any preferences as to how they want to solve their soil fertility and soil moisture problems. If so, what are they? Have they tried some techniques already? What were the results? Why?

2. How many m in elevation is the area?

3. What leguminous grains, pods or leaves do people already know and eat. In what quantities do they eat them and how do they prepare them?

4. When asked about their major agricultural problems, do farmers see soil

fertility or soil moisture content as one of the top two or three? Try not to bias the answer by former comments, how the question is asked, or even your organization's name or logo on your vehicle.

5. What are the dominant agricultural crops? Are other plants intercropped with them? If so, what percentage of the land dedicated to the dominant crops is intercropped? Do farmers use a crop rotation? What is the rotation, including the management and seasons for each crop?

6. Do farmers have grazing animals? Do people set them free during the agricultural off-seasons? Have the people set limits on where they roam? If so, who manages these limits and how? Do people commonly make fencing to keep the animals out of their fields? How much labor does fencing require, how sturdy is the fencing and how long does it last? How much space does the fencing occupy?

7. What is the average annual rainfall? During which months does it rain the most? Are there two rainy seasons each year, or only one? If two, which crops are grown in each, and why?

8. Have other agricultural programs worked in the area previously? What practices did they recommend, and what were the sustainable results of those efforts? If there were none, why? What do the farm families feel about these efforts? Do other organizations presently work in the area? If so, do they work with gm/ccs? What are their priorities regarding gm/ccs?

9. Which agricultural activities are carried out by women? What rights do women have related to crop or animal selection and use? From which farm activities, if any, do the women receive or have control of the income?

10. Have people already attempted to use natural means to improve soil fertility or soil moisture retention? If so, what were these methods? How do the farmers feel about using plants to fertilize the soil or keep it moist? Have they ever seen this done successfully? If so, what species was used? Do people use animal manure? Do they use compost? How many t/ha of each?

11. What percentage of farmers in the area still fallow land? For how many years at a time? Do farmers plant or harvest anything from their land as it is being fallowed?

12. What is the dominant land tenure system? Do farmers own the land they farm? If not, can they be sure that they will be able to continue farming any land they have improved? What is the average size of the plot(s) belonging to or rented by each household? How many households have larger holdings,

and how much land do they have? (You may want to ask these questions on land tenure of certain people you particularly trust. Normally the answer from most villagers will not be too trustworthy.)

13. Do women possess land? Do they have any rights to use land owned by others or to use trees or any other resources on other people's land? Are these use rights long-term and secure? How do women's agricultural priorities differ from those of the men? (e.g. Do women want to plant different crops than men? Do they appreciate more the value in producing firewood on the farm? Would they value more a decrease in weeding work?)

14. In general, are yields of subsistence crops increasing or decreasing each year? By how much?

Through field observations

In addition to these interviews, make careful observations. Often it is important to verify verbal information by observation, since interviews alone can lead to mistaken, incorrect or incomplete information. Observe things like average yields, the prevalence of intercropping, size of landholdings and men's and women's roles in agriculture. Observe also the following:

1. Is N a major limiting factor in the soil? (If the leaves of maize, sorghum or millet plants all turn yellow, especially in the center of the lower leaves, N is almost surely the limiting factor.)

2. Do different fields vary widely in soil fertility or moisture retention? Why?

3. Are there large areas of fallowed land? Wasteland? Forest? How old are the trees in different patches of fallowed land? Are there lots of ferns in the forested land? (Ferns usually indicate the soils are very acid.)

4. Is the agriculture quite intensive? For instance, are the crops well weeded? Is every scrap of land in use? Are technologies appropriate to intensive systems being used, such as terracing, contour ditches or barriers, contour stone lines (diguettes), zai holes, etc.?

5. What kinds of erosion are problematic? Is the land mostly hilly or steep?

6. Is there a great variation in how the farmers of the area use their land? Is a variety of crops being planted (an indication of innovativeness and/or market access)?

7. Are most of the fields' crop residues consumed by animals? Are they burned?

8. Do farmers apply to their soil large amounts of animal manure, compost and/or sweepings from their homesteads? Do they use synthetic fertilizer? To which crops do they apply it? In what amounts?

PART II: CHOOSING THE BEST GM/CC SYSTEM FOR EACH SITUATION

Chapter 8. Getting Started

Characteristics of the Recommended Gm/cc Systems

The number of different known gm/cc systems are so plentiful that the most challenging task in working with them is the selection of the most adequate gm/cc systems to use in a specific situation. For that reason, most of the second part of this book is organized in the form of a decision tree, with ample explanations of its why's and wherefore's. By following the numbers and directions given, one should be able to arrive at the anywhere from one to six gm/cc systems that would be best suited to the situation in which you are working.

Most of the 150 gm/cc systems described in this book appear in numbered text boxes. Following the number of the text box are one or two words that identify the way in which the gm/cc is used in the system, such as a "rotation" or "dispersed shade." If there is nothing to indicate the kind of system used, it is an intercrop. The species named first is the crop with which the gm/cc is associated, followed by a slash and the name of the gm/cc(s) used in the system. The remaining designations give the category the gm/cc system belongs to (e.g. "proven," "good potential," etc., as described in the preface), the countries or regions in which the system is presently being practiced most widely, as far as we know, and the approximate number of farmers who have adopted the system. A majority of the rest of the information about the systems comes from personal interviews with the farmers that my Guatemalan colleague, Gabino López Vargas, or I have conducted with the farmers while personally observing more than 100 of these systems.

The numbers of farmer adopters tend to be only rough estimates, since Gabino or I last visited some of them as much as 25 years ago. Also, a fair number of these systems are used by tens or hundreds of thousands of farmers, making it difficult to obtain more than very approximate figures.

Elevations given for the gm/cc species should be taken as a general guide, not as hard and fast rules. The best elevation for each species will vary somewhat according to the variety being used, the quality of the soil (better soils will expand the range of altitudes at which they grow well), and the distance from the equator (the further from the equator, the lower the elevation at which they will grow best).

Still, we should remember that gm/ccs must grow vigorously if they are to be of benefit, and elevation above sea level has a tremendous impact on the vigor of all gm/cc species. A gm/cc may grow outside the range of elevations mentioned, but grow so slowly that it would be better to use a different species. On the other hand, certain benefits may cause farmers to go ahead and use a gm/cc species outside of its range. For example, lablab beans normally only grow well below 1,500 m in elevation. Nevertheless, near Jesús de Otoro, Honduras, farmers working at 1,800 m grow lablab beans as a gm/cc despite their slow growth because they value so much the lablab beans' high-quality dry-season fodder.

These systems use almost entirely leguminous species. They do not include non-leguminous gm/ccs such as the oats, forage turnips (*Brassica rapa*) and sunflowers that are used widely in Brazil, the *fonio* (*Digitaria exilis*) used in West Africa, or the three-year cassava system used in Vietnam. These Brazilian species are not very useful among smallholder farmers because they usually require the addition of urea—something poorer farmers are reluctant or unable to buy. *Fonio* is not likely to be consumed anywhere outside of West Africa, and repeated cassava plantings just do not fertilize the soil that well. Besides, we have leguminous gm/cc species that supply as many, or more, benefits as those provided by these non-leguminous species.

These systems do not include some species that might seem quite promising. These include, for instance, the siratros. These species do not provide food, and they cannot fill any known useful niche for smallholder farmers that cannot be filled by gm/cc species that *do* provide food. Furthermore, we have never seen any of them being used sustainably by smallholder farmers.

The systems described also include only a few of the gm/cc systems using tree species, which are quite common around the developing world.[82] This unfortunate omission is primarily because of our own lack of experience with such species, plus the reluctance of farmers to adopt species that take so long to produce results.

Where soils are too poor or rainfall too erratic for most gm/cc species to grow well, we must often use gm/cc species that are highly resistant to poor soils and drought, such as jack beans or tephrosia. However, these hardier species often provide few additional benefits. They are only rarely edible because they usually contain serious anti-nutritional or toxic substances. These substances probably served to defend them for millennia in these difficult environments, where alternative animal fodder was presumably scarce and therefore the plants needed these natural defenses to survive.

Little attention is paid in this book to soil types. Our experience has been that soil types often have less impact on the growth of gm/cc species than one might think. Of course, certain soil factors are crucial. Many gm/ccs will suffer when soils have extremely high or low pHs or the soils are waterlogged. When the soil has a pH near or below 4.5, *Desmodium ovalifolium* may have to be used until the increased OM buffers the pH up to a little more favorable level. In Belize, *D. triflorum* has been successfully grown in soils with a pH down to 4.0. For moderate waterlogging you can plant eucalyptus trees to soak up the excess water. In more serious cases on sloping fields, farmers can often dig parallel ditches every 20 m or so across the flow of water at a ½% slope, laid out with an "A" frame, to drain off the excess water.

The First Steps in Choosing a System

Are farmers motivated to try out gm/ccs?

Farmers with enough land to be using 8-year fallows or more may not be motivated to use gm/ccs. Natural fallowing usually requires less labor, and labor is normally the limiting resource for farmers with this amount of land. Furthermore, their soils still tend to be quite fertile. Sometimes in this situation you can promote improved fallows, such as ones using lablab beans (S9) or tephrosia (S8) as a way of reducing labor demands or increasing the amount of land for farming. If lablab beans are locally consumed, S9 would be especially promising. Otherwise, gm/ccs simply may not awaken much interest.

At the other extreme, farmers who have less than about 0.25 ha, or even up to 0.5 ha of irrigated land, may have little interest in gm/ccs because their land is too valuable to free up the space for gm/ccs. If there is little or no off-farm employment, labor becomes their cheap resource. Animal manure or household waste, preferably made into compost, can become the preferred source of fertility. Often in these cases the production system has become very intensive. High-value crops may dominate, to the point that synthetic fertilizer may become profitable.

Many governments subsidize synthetic fertilizer, either to buy farmers' votes or as a form of welfare because they do not realize any reasonable alternative exists. When fertilizer is subsidized, farmers' motivation to adopt gm/ccs is greatly reduced (see S26).

Programs may be able to overcome a lack of motivation to use gm/ccs in any of three ways:

1) explain clearly to farmers the more important benefits of gm/ccs, or better,

2) establish demonstration plots run by the program or in Farmer Field Schools. Better still,

3) if smallholder farmers are using successful gm/ccs nearby, take farmer leaders on educational field trips to observe these successes.

If, after these efforts, farmers still have no motivation to try out gm/ccs, you may need to give up on the idea for the time being.

Later, farmers' priorities may change. Your program may discover a more advantageous gm/cc system, or it may, through its work, gain more credibility. In some cases, the end of government subsidies or the corruption of the subsidies to the point that many farmers no longer receive them, will motivate farmers to try out gm/ccs. Also, over time, farmers' soils may become less fertile and more affected by drought, farms may become smaller or fertilizer prices may rise, causing farmers to become more interested in gm/ccs. In such cases, renewing your efforts with gm/ccs could be very valuable for the farmers.

Is there a successful gm/cc system nearby?

Does a successful gm/cc system exist near the program location or in an area of roughly similar farming systems and ecological environment? Do the farmers know of such a system, and are they interested in it? Are more smallholder farmers adopting this gm/cc system than abandoning it? Are they adopting it even in the absence of any artificial subsidies or outside promotion?

If a system exists and the answers are positive, read on. If one exists, but the above answers are mostly negative, interview the farmers that are using it. Be careful not to simply take their answers at face value. Analyze the system's economics and its pros and cons, both according to the farmers and according to your own analysis. A few of your best farmer leaders might participate in this analysis. If the system seems to provide more benefits than costs, try doing some field trials. If it is not beneficial, move on to Chapter 9.

Field trials

Any time that you have selected one or more gm/cc systems that are worth trying out, return to this point and carry on again from here. You should organize at least three or four trials of the gm/cc system in question. Preferably, do the trials on the land of your best farmer leaders, or of those farmers who will do the best job of managing the experiments. If possible, their farms should be easily accessible to visitors. Be careful to choose farmers who have a record of being honest and have an interest in gm/ccs. If no such option looks good, either conduct the trials on Farmer Field School plots partially managed by program personnel, or on a small plot of land managed by the program.

Test the germination rates of all seeds to be tried by placing 50 to 100 seeds between about 30 pages of a moistened newspaper. Keep the newspapers moist throughout the test period. After one to two weeks, the germination rate of most seeds should be obvious. However, with some of the more drought-resistant species, such as mucuna or jack bean, up to 40% of the seeds will not germinate until after a month's time.

These trials should always be compared to a comparable field, or control plot, where the best practices of traditional farming are being used. Keep close records of the costs and benefits of each system, being especially careful to include all labor costs. Far too often those of us working in agriculture recommend technologies without knowing for sure whether their benefits exceed their costs. If your experiments fail, check out other possible gm/cc systems in Chapters 10 to 12. If there are none, consider not working with gm/ccs. If the trials are looking successful, start organizing educational field trips to see either the fields of any farmers already using the system nearby, or the trial plots themselves. Successful experimental plots easily can and should be given a second important role—that of demonstration plots. Organize these educational field trips as soon as the results clearly look positive.

Educational field trips

Educational field trips should allow influential farmers, both men and women, to see the results of the gm/cc experiments. If the plots to be visited are within the program area and fairly accessible, take as many farmers as possible to see them. If the plots are located at some distance, try to take at least two or three respected leaders (at least one of them a woman) from each program village.

At the plots, if at all possible, have the farmer experimenters themselves present and explain the results. Give the visiting farmers plenty of time to ask all the questions they want.

Encourage farmers to take note of each of the advantages of the gm/cc system they are observing. These advantages may include weed control and high-protein food production by the gm/cc, as well as long-term increased yields and drought resistance. Two educational field trips may be required to observe all these advantages: one when the gm/cc is maturing, and the other when the next season's crops are showing the impact of the gm/cc.

If farmers are still skeptical of gm/ccs after the visits, either give up for the time being or try to find some better systems in Chapters 10 to 12. If at least some farmers seem motivated, perhaps you can start an extension effort with these farmers. If nearly all the farmers now seem interested, integrate the gm/ccs into your extension program.

Extension work

One should always start promoting the system by teaching farmers to test it on plots anywhere from 10 m x 10 m to 25 m x 25 m in size. Use a participatory extension methodology, in which farmers first experiment with the new technology (known as participatory technology development, or PTD). Once their own experiments are successful, encourage them to teach it to each other (farmer-to-farmer extension).[83]

It is very important to get as much feedback as possible on the farmer experiments. Have frequent conversations with the farmer-experimenters. If problems arise, try to trouble shoot them very quickly. Watch closely over time whether experimenting farmers use the system on larger plots each year, whether those not in the program adopt it, and whether those who have tried it abandon it. If these indicators show farmers to be positive about the gm/cc system they are using, continue promoting it and look in later chapters for other systems that might complement it.

Chapter 9. Conserving the Soil

Photo 24. Honduras. Micro-terraces like this can be easily built for an average of 1/10 the cost of bench terraces.

Soil degradation can reach almost unbelievable extremes. Farmers themselves are often the first to believe that things cannot possibly get too much worse than they already are. But they can, almost always. Severe degradation is no respecter of continents or regions. Simple water erosion can make an absolute moonscape out of land that was forested or farmed just a decade or two before (see photo 25). Agricultural burning eventually creates wastelands. Overgrazing can turn pastures into deserts. The animals may continue to graze, but the "pasture" is so meager you cannot see anything green without getting down on your hands and knees. Hillsides can become piles of rocks or moonscapes, or even just small patches of soil in among solid rocks, as in photo 27.

Photo 25. Mexico. Just one generation ago, this land was full of wheat fields. Much of the north of Oaxaca State looks like this now.

This land could be recovered, of course, but the labor costs would be immense. That labor is better invested in making present farmland more productive.

One very effective way of motivating farmers to conserve their soils is to show them photos such as these of situations where things have gotten much worse than where they live. I often use such a set of photos. It starts with photos of forests being burned off, then fields gradually losing their fertility, and finally, wastelands and deserts far worse than anything they have ever seen. For many of these farmers, this was the first time they realized that if they continued to do what they were doing, their situation and that of their children could be a whole lot worse. It was a prospect they dearly wished to avoid.

Photo 26. El Salvador. To increase gains in sustainable productivity, we need to maximize biomass production. Fires destroy biomass. As long as burning continues, the land must either be fallowed, very expensively fertilized, or just allowed to gradually deteriorate until it becomes a wasteland.

If nearly all the farmland in your program area has a 7% slope or less, or if most fields have less than a 15% slope and gm/cc systems will keep the soil well-covered either with a green or dry mulch, move on to Chapter 10. If neither condition applies, continue here.

Photo 27. Vietnam. This woman is hoeing her field. Soil once completely covered these rocks, but now the fields consist of tiny pockets of soil held in among the rocks. Centuries of erosion have left almost nothing to farm. Not even months and months of work could restore these fields.

A 15% slope may seem like it will suffer from erosion, even with cover. If you have any doubts under your conditions of rainfall and soil type, apply a 10- to 20-cm thick layer of mulch over a 1 m X 1 m piece of land in a field with a 15% slope. Dig a small ditch above your experiment to protect it from water flowing down the hillside. Then check to see if during a heavy rain, any water flows out of the plot, and whether that water is clear or is carrying away soil.

Most smallholder farmers make their own amazingly accurate calculations as to how much a soil conservation technique will cost and how much it will pay in increased income. Farmers with access to irrigation and good markets, who produce high-value cash crops, like vegetables or rice, will often build relatively expensive structures like micro-terraces. However, farmers who grow inexpensive subsistence crops in remote areas will only be willing to use simple, inexpensive techniques, like contour grass barriers. Farmers having more land, and those having less security of tenure, will also tend to invest less effort/ha on soil conservation.

All soil conservation technologies require maintenance—some of them quite frequently. If farmers fail to do adequate maintenance, these technologies often result in worse erosion than if they never existed, because they channel the water into more concentrated flows. Somewhat benign rill erosion becomes more damaging gully erosion. Therefore, farmers themselves need to have invested enough in soil conservation to maintain it. Payment by a program for soil conservation work, whether with cash or food for work, is usually counter-productive: farmers build the structures to receive the payment, but seldom maintain them once the payments end. I have seen thousands of hillsides where subsidized soil conservation work has had a negative impact (see photo 30). On several occasions, I have also known farmers to pull down contour rock walls they were paid to build, so they could earn more income from a program by rebuilding the walls the next year.

The Relationship Between Soil Conservation and Soil Restoration

The very first thing water will carry away is the OM on top of the soil. If gm/ccs are to have any impact, the flow of rainwater down a hillside must be slowed or stopped. Once a gm/cc starts building up SOM levels, the soil becomes increasingly soft and spongy, so more and more rainwater will infiltrate into the soil (see photo 7). Thus, over a period of several years, the need for maintenance on soil conservation works diminishes. In northern Honduras, for instance, where the 3rd maize/mucuna system (S52) has been in use for over 50 years, the topsoil acts so much like compost that no water runs down hillsides with a 30% slope, even though the annual rainfall surpasses 2,000 mm/year and no soil conservation measures exist.

In addition to preventing erosion and increasing rainwater infiltration, gm/ccs are tremendously important on hillsides for another major reason: their ability to increase productivity. Many people think that soil conservation measures will improve productivity on hillsides. However, research shows that while soil conservation will increase productivity just above the contour barriers or terrace faces, this increase comes at the expense of an equal drop in fertility just below the barriers or terrace walls. That is, the increased yields above the structures are merely a case of robbing Peter to pay Paul.

Soil conservation can keep soil on a hillside, but it does not improve soil fertility. For that, we need soil restoration—most efficiently with gm/ccs. In the end, to achieve better productivity on steep hillsides, we need both soil conservation *and* soil restoration.

Ground Cover and Contour Barriers

Entire books have been written about soil conservation. Here, I will briefly cover only the aspects of soil conservation that are most important, least well-known and, in a few cases, most controversial. For instructions on the technical details of these technologies, please consult books or training manuals dedicated to soil conservation.

Ground cover

A good ground cover during the whole rainy season, consisting of either a living "green mulch" or of dead OM at least 5 cm thick, is far more important than most people think. The vast majority of erosion occurs because raindrops falling from the sky hit the soil with such impact that they break up soil aggregates into small fragments that are easily carried away by running water. Mulches of either kind break the fall of raindrops, making them largely harmless. Over time, ground cover can often contribute more to preventing water erosion than do physical barriers, whether these be earthen bunds, rock walls or ditches. Such a ground cover will be an automatic by-product of most gm/cc systems.

Contour hedgerows

Photo 28. Haiti. The A-frame is the indispensable tool that allows smallholder farmers to do soil conservation. It in no way needs to be fancy. The poles do not have to be straight, nor the joints held by nails. The main requirements to achieve accuracy are that the joints be held together firmly, and the poles are stiff (i.e. inflexible). In areas of mostly flat land, like West Africa, a water level is usually preferable.

Contour vegetative hedgerows usually prove more advantageous than non-vegetative barriers because they require much less labor to establish and keep up, and they may also produce food or fodder. Grasses normally work best. Non-grassy species in hedgerows (such as trees, pineapples, etc.) will seldom hold a terrace face of more than 0.4 m, whereas grasses can easily hold a terrace face of up to 1.3 m in height (see photos 31 and 32).[84] If farmers nevertheless wish to plant non-grass crops in their hedgerows, a good option is to first plant the grass and then space the non-grass crops along the grass barrier. If farmers want to use dispersed shade on hillsides, they can plant every second or third row of dispersed trees on a contour right along their grass hedgerows.

The best over-all species for hedgerows is dwarf Napier grass (a variety of *Pennisetum purpureum* that grows shorter in stature than most Napier grass and has almost no fuzz, or pubescence, on the leaves). Compared to other Napier grasses, this variety is much more palatable to animals, competes less with nearby crops, and does not invade farmers' fields as quickly. Dwarf Napier grass produces huge amounts of fodder: less than 100 m of hedgerow can permanently feed one milk cow all it can eat all year long. If free-grazing animals are eating all the fodder so the owner's animals do not benefit from it, vetiver grass (*Chrysopogon zizanioides*) is usually a preferable choice.

Photo 29. Honduras. Where the weather is hot enough, sugarcane barriers are a very popular species because the farm family can either use the sugar or sell it. Sugarcane is also an extremely nutritious dry-season fodder for cattle.

If a field has a slope of 12% or less, it is less than 800 m in elevation and grazing animals are under control, sugarcane can be a very popular hedgerow species. It should, however, be avoided on steeper slopes, since it tends to fall over if the terrace face grows more than 30 cm in height. Plants such as lemon grass (*Cymbopogon* spp.) for making tea can be used instead of Napier grass on appropriate sections of people's hedgerows. However, these plants also cannot hold a terrace face of more than 50 cm in height.

In vegetable gardens, to reduce the width of a hedgerow and the amount of shade it casts, short-stature grasses like weeping lovegrass (*Eragrostis curvula*) work much better, assuming once again the terrace faces will remain less than 30 cm in height.

Photo 30. Haiti. All soil conservation measures require at least some maintenance. These rock walls once covered this hillside, but an NGO paid the farmers to build them, so no maintenance has been done since the payments stopped. As a result, nothing worthwhile remains in the field. Even the soil has been washed away.

Hedgerows should not be planted less than 12 m apart, because farmers usually reject any closer spacing. If hedgerows 12 m apart are too far apart to catch all the soil, do not move them closer together; instead, find a way to cover the soil better with a gm/cc. The rows can be as far as 20 m apart on less inclined slopes.

Photo 31. Honduras. This dwarf Napier grass hedgerow has caught 1.7 vertical m of soil in its six or seven years in existence—soil that otherwise would have been lost to this farmer. (The farmer trimmed the hedgerow lower than usual so we could more easily appreciate the height of the terrace it has created.)

Photo 32. Mexico. On a slope that is less steep than in photo 31, Napier grass has held tons of soil on the slope, gradually forming a virtually flat terrace. The farmer did not have to move any soil at all; rainwater did all that work for him. A small ditch at the foot of the hedgerow keeps any water flowing through the barrier from causing problems.

Photo 33. Peru. The Wari people built these stone walled terraces 500 to 1,000 years ago, before the rise of the Incas. The walls are still in use—which means they have been maintained all that time. Contour hedgerows cost perhaps a tenth of the work required to make rock walls.

Hedgerows should always be laid out using an A-frame level. Even when farmers feel confident that they can make level contour lines by eyeing the slope, the contours will not be accurate. As a result, rainwater will accumulate at the lowest points in the contour lines, and break through.

Where waterlogging on hillsides is causing problems, use the A-frame to orient the hedgerows at a 0.5% slope across the hillside. Then dig small drainage ditches just below the barriers, ending, if possible, at areas on the slope of natural drainage, so the excess water can continue down the hillside.

Unless farmers need the terraces right away (e.g. a new business nearby has suddenly provided a high-paying market for quality vegetables), contour hedgerows are to be favored over building bench terraces because the hedgerows gradually build bench terraces, with no additional labor required once the grass has been planted (see photos 31 and 32). In South America, these terraces built by hedgerows are called "slow-forming terraces."

Farmers should plant their hedgerows before they start using gm/ccs.

Photo 34. Honduras. This contour rock wall has saved a 1 m layer of soil from washing down the hillside. Just as the Wari and Inca people did, modern Honduran farmers place the larger rocks facing outward and the smaller ones in the interior of the wall.

Contour rock walls

Given the lower cost of hedgerows for making terraces, contour rock walls are only advisable when the rocks in a field are making farm operations difficult. Digging up buried rocks or transporting them from other

Photos 35 and 36. Photo 36 by Laureano Jacobo. Honduras. The two photos on this page of the same field, shot about 10 years apart, show some of the problems one faces with rock walls. First, in the second photo, the farmer obviously had to build the walls taller, because the soil had built up to the top of the smaller walls. Second, even with the extra work, the middle wall is resting on parent material, without any protection from the soil on the terrace below. Eventually, the base of this rock wall could erode away.

Nevertheless, the results are positive enough that the farmer of the field in the background just beneath the road, who had still been burning his field in the first photo, built his own rock walls (visible in the second photo). This second farmer's rock walls are not on a strict contour showing they were not laid out using an A-frame. He was therefore probably a spontaneous adopter, not actually affiliated with the program.

fields is almost never worthwhile. If the surface rocks in the field do not build a tall enough barrier, or in time the soil builds up to the top of the rock barrier, a strip of dwarf Napier grass or vetiver can be planted just 2 or 3 cm uphill from the rock barrier. That will gradually raise the combined rock-and-grass barrier even higher as more soil builds up behind the grass.

Strip Tillage and Micro-terraces

Micro-terraces are an attractive option for farmers near markets who want to plant vegetables or other potentially high-value crops on land with slopes between 15% and 60%. As the land gets steeper, the contour tilled strips become thinner and taller, changing from strip tillage into raised beds and eventually micro-terraces. (Compare the photos of strip tillage on increasingly steep land, starting with photo 37 and moving through photos 39, 24, 38 and 40, in that order.)

Photo 37. Honduras. A farmer and his son prepare a piece of strip tillage for the following season. This work is only needed the first year or two, until the OM within the row becomes plentiful enough for farmers to convert to no-till. The farmer has placed all the rocks in the grassy strip between the rows, where they will no longer cause any problems.

On land with more than a 30% slope, tilled strips become micro-terraces over about three years. These micro-terraces therefore only take about 1/10 as much labor as that required to construct bench terraces, although they do require a fair amount of maintenance.

Photo 38. Honduras. These micro-terraces hold almost all the rainfall. Virtually no erosion occurs because the water that does run off the micro-terraces flows along between them at only a 0.5 % slope.

On the vertical micro-terrace faces, the weeds should not be hoed or tilled, but merely cut off slightly above ground level, so the indigenous grasses will hold the soil on the terrace faces. With this technology, farmers often prefer steeper slopes because as the terrace faces get taller, "stoop" labor is eliminated; farmers can almost stand up straight as they work on their vegetables.

Photo 39. Honduras. Where the land is not very steep, micro-terraces can be made fairly wide, so they look and function like raised beds in a vegetable garden. A small irrigation channel can run along the uphill side of each micro-terrace.

Furthermore, on the steeper slopes, the micro-terrace face works as a water harvesting surface, concentrating the rainfall on top of the micro-terrace below. Once again, the micro-terraces are laid out in contours with an A-frame level, with a 0.5% slope, so excess water will drain away slowly in one direction.

The micro-terraces in photo 40 look quite fragile, but are remarkably resilient. In Honduras during Hurricane Mitch, all the rainfall that usually falls in a whole year fell in four days. Micro-terraces like these, including the field in the photo, suffered only minimal damage. Some nearby hillsides that were being fallowed, with roughly the same slope and no soil conservation, slid down the hill.

The main drawback with micro-terraces is that they require more work, to build as well as to maintain, than the other soil conservation measures recommended here. For this reason, farmers will only sustainably use them

Photo 40. Honduras. Micro-terraces work well on slopes up to 60%. This field and others like it produce some of Honduras' finest vegetables. Although a few terraces in the foreground are staked because the soil is sandy, native grasses alone are usually able to hold the terraces. Here, carrots are growing in the lower foreground, coffee in the background.

if they achieve a good income. Experience indicates that only farmers growing high-value crops, often with irrigation, and located within three or four hours' drive of a good market, will use them sustainably. Also, since micro-terraces stop all soil from washing down the hillside, they do not allow "slow-forming terraces" to form behind contour grass barriers.

Diguettes

Diguettes are usually used on land with less than a 3% slope, such as in most of the Sahel. Diguettes are lines of rocks, usually just one or two deep, that run on a contour across waterways (called "wadis") that only have water in them after a good rain. Diguettes function much like larger rock barriers on a steeper hillside: they slow the water flow (even though the water flows through them), gradually build a small terrace, cause more infiltration of water along the diguette and capture some of the more fertile, eroding soil.

Used widely in several Sahelian countries, diguettes have made a major contribution to increasing soil moisture retention and reducing the loss of fertile soil. Sometimes farmers also plant grass barriers above them, similar to what is done with rock walls when the soil fills up behind them.

Zai holes, planting pits and half-moons

Zai holes, planting stations and half-moons are depressions dug into the soil surface to catch rainwater that is flowing across the surface, as well as to place OM or fertilizer and seeds. These three technologies basically differ only in size and shape. Zai holes are normally round and 20 to 25 cm in diameter; planting pits or planting stations can be anything up to 50 cm square; and half-moons are usually 3 to 5 m in length and shaped as their name implies.

Photo 41. Burkina Faso. The millet in the foreground in northern Burkina Faso was planted in the traditional manner. The millet in the background was planted with zai holes containing a handful of animal manure. By catching rainwater, zai holes concentrate both moisture and fertility near the crops' roots.

Farmers rarely make half-moons sustainably or in quantity because they require so much work. Even planting pits, as small as they are, seldom last long after a program ends. When they do, farmers seldom maintain them on more than ½ ha.

During an evaluation I carried out eight years after the end of a program in Mali, farmers reported that zai holes with manure in them increased productivity enough to pay for their labor, but zai holes alone did not. As a result, they continued maintaining only the number of zai holes needed to use all of their manure.

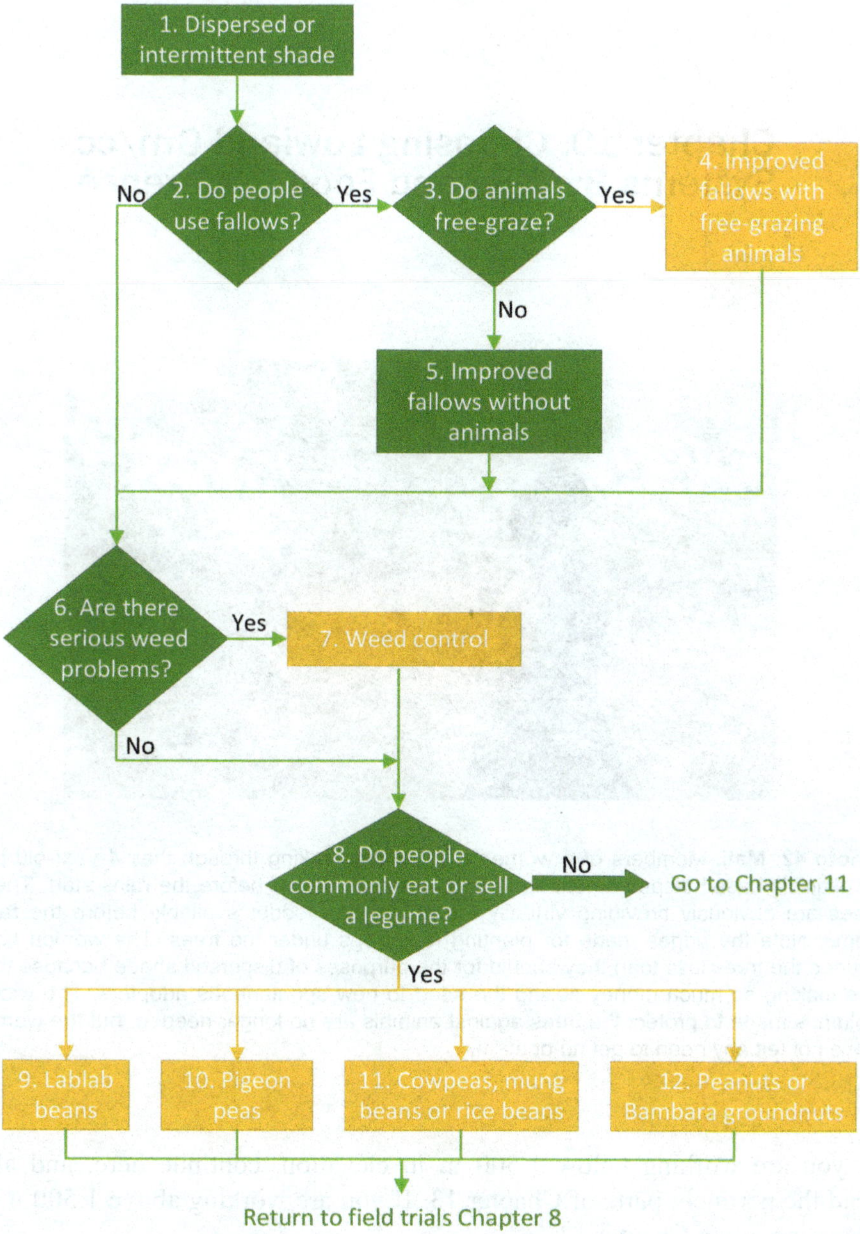

Numbers correspond to numbered commentary sections in Chapter 10

Chapter 10. Choosing Lowland Gm/cc Systems By Type and Food Preference

Photo 42. Mali. Members of a women's group are walking through their 4-year-old plot of gliricidia-based dispersed shade in April, a month or two before the rains start. These trees are obviously providing virtually the only green fodder available before the rains come. Note the ridges made for planting row crops under the trees. The women have pruned the trees less than they should for the purposes of dispersed shade because they are making so much money selling the seed to new spontaneous adopters. The wicker columns made to protect the trees against animals are no longer needed, but the women have not felt any need to get rid of them.

If you are working below 1,500 m in elevation, continue here, and also read the pertinent parts of Chapter 13. If you are working above 1,500 m in elevation, go to Chapter 12.

Dispersed or Intermittent Shade

1. Dispersed or intermittent shade. These systems can and should be used virtually anywhere below 1,500 m in elevation. At the elevations approaching 1,500 m, a lighter shade (perhaps 15 to 25%) will be more appropriate, although the pruning can usually be done the same as at lower elevations, since most of these tree species grow more slowly at higher elevations. Dispersed shade systems are described in S3 to S7. An intermittent shade system is described in S67.

The Advantages of Dispersed or Intermittent Trees

The primary advantage of dispersed shade is that crops beneath them grow better. In the lowland tropics, the mid-day heat becomes so intense that virtually all unshaded crops stop growing for up to three hours.[85] A light shade can increase crop yields by as much as 30%. And as global warming intensifies over time, the trees can be pruned less and less, thereby stabilizing the temperature where the crops are growing, while the environment, in general, becomes hotter.

Scientific experiments done in Honduras and Australia have shown that an average shade of between 15% and 50% will increase the growing period of crops by about two weeks after a good rain.[86] This occurs because the lower temperature under the trees reduces both the evaporation and transpiration rates, while the trees themselves reduce the wind speed. All three factors improve soil moisture retention.

By both decreasing ambient temperatures and maintaining soil moisture longer, dispersed shade largely solves the two factors of global warming that most affect agriculture: higher temperatures and irregular rains. Dispersed and intermittent shade thereby provide by far the cheapest long-term defense farmers have against global warming.

The products that grow on the dispersed trees themselves add even more to the income and benefits coming from the system. These products can include edible fruit, flowers or leaves, fodder, timber, firewood, natural pesticides, gm/cc biomass and more. Additionally, year-round roots in the soil increase the populations of beneficial soil microorganisms, especially fungi and mycorrhizae.

The most universally overlooked advantage of dispersed shade is that it reduces the loss of N caused by volatilization. By lowering the ambient temperature by about 10° C., dispersed shade reduces N volatilization enough that it is no longer a problem. Furthermore, since the annual pruning is done at the end of the dry season, the amount of shade during the dry season, and consequently the reduction in volatilization, will be even greater while crop residues are lying in the fields. Leguminous crops such as peanuts, cowpeas, mung beans, rice beans and perennial mucuna are all suddenly transformed into useful gm/ccs anywhere in the tropical lowlands. In the long run, this advantage could well become the single most important benefit of dispersed and intermittent shade.

All these advantages come with only minor labor costs: planting the trees, protecting them one or two years from grazing animals and termites, and pruning them once a year (or perhaps twice a year in areas of bimodal rains). Even the pruning reduces net labor costs: it takes a lot less work to prune trees in one's own field than having to climb trees and cut firewood in a forest and then carry it all home. In addition, growing one's own firewood saves forests.

All these many advantages should not surprise us. After all, trees in farmers' fields move farmers one giant step closer to imitating one of nature's most productive and sustainable ecosystems: the tropical forest.

Traditional dispersed shade systems already exist in many areas of the developing world. One of the most notable cases is the millet/*Faidherbia* system (S5) used across the northern Sahel from Senegal to Chad. Traditional systems are also to be found in a dozen nations of Latin America—most notably Mexico, El Salvador, Guatemala, Honduras, Nicaragua and Brazil.

Nevertheless, we professionals have only recently begun paying much attention to this technology. Probably the best single study ever done of traditional dispersed shade systems was carried out by Malcolm Cairns, who worked in Southeast Asia with ICRAF. He spent years during the 80's and 90's cataloguing dozens of dispersed shade systems spread from northeastern India to Myanmar and China.[87] As far as I know, Dr. Ian Cherret and Ing. Luís Alvarez Wélchez of the FAO, carried out the first scientific experiments with dispersed shade, based on the traditional "quesungual" system in southern Honduras.

Most dispersed shade systems consist of trees planted 5 to 8 m apart in rows, with 5 to 8 m between the rows, creating a population of between 140 and 400 trees/ha. If farmers are using bushes (for example, pigeon peas or tephrosia to shade low-growing vegetables), the spacing might be only 2 to 2.5 m apart.

Photo 43. Mexico. These young gliricidia trees are serving both as dispersed shade and, in just three or four cases until the gliricidia grows larger, as tutors for vanilla.

The cover crop is perennial peanut (*Arachis pintoi*), one of the world's finest cover crops. It will provide additional fertility to that of the gliricidia. This field has not been weeded in four years, and yet it has no weeds. Furthermore, the perennial peanut does not climb the trunks of the trees.

At least some biodiversity among the trees is always desirable in a dispersed shade system. This way, if one tree species were to die out, others would remain. Often, if desirable local trees already exist on the land, we leave those that already happen to be growing in the rows where the dispersed trees will be planted. Nevertheless, we must keep in mind that the native trees, when grown from seed or seedlings, will often grow more slowly than does gliricidia.

In some areas of the world, including parts of Honduras, Mali and the Philippines, farmers prune the trees in a natural forest in order to grow crops beneath them. This system has major advantages in terms of reduced labor costs, biodiversity and drought resistance, but may not have enough N-fixing trees to help with soil fertility. Nevertheless, we need more experience with such systems. Certainly, the addition of some gliricidia or leucaena trees in the forest would help solve any N problem that might exist.

Disadvantages

All trees grow rather slowly. Programs often want quick, recognizable results to maintain farmer enthusiasm. Therefore, you might want to delay planting dispersed trees until after you have had some success with other agricultural technologies first. If so, go to the heading "Improved Fallows" in this chapter.

Another disadvantage of dispersed tree systems is that because the trees are planted so sparsely, these systems require additional gm/ccs if they are to improve soil fertility over the long run.

Dispersed Shade Species

Gliricidia. Gliricidia is the best tree we know of for dispersed shade systems. It fertilizes the soil far better than any other trees except, possibly, leucaena or *Piliostigma.* Under Sahelian conditions, gliricidia grows from seed three or four times faster than *Faidherbia* or *Piliostigma*, and it is not invasive, like leucaena.

S3. Dispersed shade/gliricidia/*Piliostigma*. Good Potential. Mali. 6,000 farmers. Appropriate for areas with more than 400 mm of rain, this system includes gliricidia trees as dispersed shade with hedgerows of *Piliostigma* along the rows of gliricidia. Space the gliricidia trees with 5 to 7 m between rows and within each row. Rows of *Piliostigma* are planted along every second or third row of trees, with 1 m between individual plants. In the drier northern areas, where the *Piliostigma* grows rather slowly, tephrosia can be planted along with the *Piliostigma* to provide plenty of OM and nitrogen until the *Piliostigma* is large enough to do so.

Even with just five months of rain a year and on depleted soils, the gliricidia trees and the *Piliostigma* together produce enough biomass within four to seven years to increase soil fertility. If the system includes any traditional cowpeas, peanuts or bambara groundnuts, the system should increase soil fertility and soil moisture retention quite rapidly.

For many farmers, the most important benefit of this system is that it provides plenty of gliricidia fodder during April and May, the last two months of the dry season. This is precisely the period when many cattle in the Sahel die for lack of fodder if the previous rainy season was droughty—which nowadays happens about one year in three. Social prohibitions in the Sahel against cutting other people's trees mean that a farmer can be reasonably sure the fodder will be available for his or her animals, not someone else's. More than 6,000 women use this system in some 90 villages of Mali. They have been selling gliricidia seed to other farmers at US $ 5.00/kg—an incredibly high price for gm/cc seed—largely because of the trees' reputation for saving people's cattle. Not only are the women making good money, but their enterprise is spreading the gliricidia far and wide.

The story gets even better. The leaves have a protein content of over 20%, yet their palatability is low, so cattle will not eat them during the rainy season, precisely when their N is needed for fertilizing people's crops.

Farmers should allow native trees to grow in this system, too, as long as they grow within the rows of gliricidia. A few moringa (*Moringa oleifera*), another promising species, could be included and pruned the same way as the gliricidia. Its green leaves can be eaten, both cooked and raw.

The primary problem with this system is that in drier areas (below about 600 mm of rain), farmers will need to guard young trees and bushes from free-ranging grazing animals during the first and second dry seasons. Farmers should circle the gliricidia trees with sturdy wicker defenses having a diameter of about 1.5 m, made with the branches of local thorn bushes (see photo 42). They can protect the *Piliostigma* bushes the first year just by placing the branches of the thorn bushes over the young *Piliostigma* plants. By the second year, animals will not bother the *Piliostigma*.

S4. Dispersed shade/gliricidia. Proven. Honduras and El Salvador. Tens of thousands. The original gliricidia-based dispersed shade system, called "quesungual," was developed by smallholder farmers in southern Honduras and eastern El Salvador, where gliricidia is native. Through the years, they allowed gliricidia trees that volunteered and were well-spaced to grow in their fields. In this way, they eventually had fields with gliricidia trees spaced at an average of about 8 m distance. As a result, soil fertility decreased much more slowly over time.

Managing the trees' shade takes very little work. A farmer can cut off the stem just above head height, and every year after that, repeat the operation. The tree will sprout again each time (see photo 19). Farmers usually cut off all the branches each year. However, where termites are common in Africa, three or four of the smaller branches should be left in place each year so that more of the trunk will stay alive. Since termites prefer dead wood over live wood, this simple procedure will prevent termite attacks in most places.

Faidherbia. In areas with less than 400 mm rainfall/year, *Faidherbia*-based dispersed shade (S5) is another popular system. Because it stays dormant during the rainy season, it does not compete with nearby rainy season crops for water or nutrients. Therefore, in spite of its thorns and the lack of N in leaves that fall naturally rather than being pruned, *Faidherbia* continues to be the non-food producing species that farmers in the northern Sahel prefer to leave in their fields.

Nevertheless, as one moves to areas of higher rainfall, the advantages of *F. albida* diminish in importance, and the problems of its thorns and slow growth make it less desirable than several other species.

S5. Dispersed shade/*Faidherbia*. Proven. All across the Sahel. Hundreds of thousands. Farmers purposely leave existing *Faidherbia* trees in their fields. Unlike most trees, *Faidherbia* trees have no leaves during the rainy season, therefore supplying a very beneficial 15 to 25% shade for the crops beneath. In this way, they are unique in creating a natural dispersed shade system that requires no pruning

Leucaena. Because of its tendency to become a pest, the leucaena tree would probably only be successful as a dispersed shade if the landholdings are quite small, so there is plenty of labor to keep new sprouts from becoming trees.

If dispersed shade is needed for vegetables, see S6 and S7.

S6. Dispersed shade for vegetables/pigeon pea. Good potential. Sudan. Tens of thousands. Pigeon peas can be used to supply dispersed shade to any vegetables that grow less than about 1.3 m in height. They offer several advantages over other species used for this purpose: their roots bother tilling operations less, they are much more flexible when garden spacing changes, they are easy to trim, they provide plenty of shade within 4 months of planting and they provide a high-protein grain for people or animals. To provide dispersed shade, farmers can plant the pigeon peas about 2 m apart each way. They will need to replant the pigeon peas every four years. Any time the shade becomes too great, they can easily trim, ratoon, or pull out some of the pigeon pea plants.

S7. Dispersed shade for vegetables/tephrosia. Fallback. Honduras. Very few. Tephrosia can be used as a gm/cc and dispersed shade for vegetables in the same manner as pigeon peas (S6). Yet since tephrosia produces no food, this system would only be advisable if, for some reason, pigeon peas cannot be used.

If a dispersed shade system, traditional or introduced, already exists nearby, take farmers to see it. If no such system exists nearby, it is usually impractical to do field trials of this technology because these species take so long to mature. Instead, together with a group of interested farmers, select a species with which to experiment. These experiments can also serve as a source of cuttings or seed for future dissemination.

Improved Fallows

In almost any improved fallow system, the soil will be very poor when the gm/cc is planted. Farmers using improved fallows will therefore have to use gm/cc species that grow well in heavily depleted soils. These would include jack beans, tephrosia, pigeon peas, *Piliostigma* and certain trees like *Faidherbia* and gliricidia.

2. Do fallows still exist in the area? If nearly all the farmers in the area still use a fallow, even if just for two years, read on. If most farmers have insufficient land to fallow, they will not be able to use improved fallows. In this case, programs would best avoid promoting improved fallows, because the poorest families would be left out. Instead, skip on to the noxious weeds section below.

Improved fallows are the simplest way of introducing a gm/cc into a program area. They are generally quite popular with farmers because they reduce labor and increase the land available for cropping. Planting and cutting down a gm/cc requires a lot less labor than cutting down a forest. An improved fallow will also reduce the fallowing period from today's two to six years down to only one or two years. As a result, a farmer can grow crops on the same land four to eight years out of every decade instead of just two to four years each decade. Farmers immediately understand the benefit of this change: the land available to them for cropping suddenly doubles!

3. Do grazing animals roam free in the dry season? If animals in your area roam free during the dry season, read on. If not, move down to "Improved fallows where the animals are kept away from the fields."

4. Improved fallows where animals free-graze. In this situation, your best approach is to test the following four species to determine which of them controls the weeds best and whether the animals get hungry enough to eat them. These species are climbing jack beans (or bushy-type jack beans if the climbing ones are not available), sword beans, *Piliostigma* and tephrosia (either *Tephrosia vogelii* or *T. candida*). Grazing animals rarely eat any of these species in areas with more than 600 mm of rain a year. Below that level of rainfall, however, animals may eat any of them before the end of the dry season.

A week or two before the first rainy season of the fallow period begins, broadcast the seed of each gm/cc variety in a test plot. Sow the jack bean or sword bean seeds at a rate of about three seeds/sq m and the tephrosias and *Piliostigma* about one to two seeds/sq m. Read S8 to see how such a system is used in Cameroon.

Photo 44. Cameroon. This tephrosia-based improved fallow system (S8) spread from one farmer to over 2,000 farmers within 15 years, as farmers spontaneously taught each other how to do it. This example, along with many others, shows the simplicity and innate popularity of simple, carefully selected gm/cc systems.

If the livestock eat all these species, you will have to either convince farmers to control their grazing animals or give up on using an improved fallow. If you tested the seeds during a good year for rain, you may have to run the test again during a year of poor rains.

S8. Improved fallow/tephrosia. Proven. Cameroon, Thailand and Vietnam. Over 2,000. Farmers near Bamenda in Cameroon use tephrosia in an improved fallow system. They broadcast tephrosia seeds onto a field before the rains begin in the first year of fallow. One year later, they cut down the tephrosia and crop the field again. In this way, the soil becomes fertile enough to farm again in just one year, instead of four years, as is normal under a natural fallow in this area.

According to the traditional system, land was cropped for three years and then usually fallowed again for four years with a natural fallow.

The tephrosia fallow, then, allows a farmer to crop the land three out of every four years instead of three out of every seven, increasing the amount of land being cultivated in any given year by about 75%. Thousands of farmers in Vietnam also use this system, which apparently spread spontaneously there also.

Even better, farmers could intercrop tephrosia with the maize or cassava crop that is planted before the fallow starts. The tephrosia grows slowly while shaded by the maize or cassava, but grows well once these crops are harvested, covering the field for much of the dry season. Used regularly, this system should prevent ever having to fallow the land at all.

S9. Improved fallow/lablab bean or jack bean. Experimental. Mozambique and Madagascar. Very few. As is done with tephrosia in S8, farmers can use lablab beans or jack beans to shorten longer fallows to just one year. Once again, they could also intercrop either of these species (using bushy-type jack beans) with the last crop of the rotation before the fallow period, thereby eliminating the fallow altogether.

S10. Maize/*Piliostigma*. Fallback. Ghana. Tens of thousands in the past. Near Bolgatanga, just after farmers have cut down the fallow trees on their land but before they burn the felled trees, they cover all the *Piliostigma* bushes with soil. Their ancestors must have realized that when these bushes were burnt, they no longer fertilized the soil nearly as well. The significant added effort put into protecting the *Piliostigma* leaves from the fire is an important testimony to the leaves' ability to fertilize the soil. Traditionally, this system was used with all local crops, but now it is disappearing because fallowing itself is disappearing.

This system should only be recommended if farmers are still burning and no other system seems possible.

5. Improved fallows where the animals are kept away from the fields. If grazing animals are kept out of the fields, it is much easier to use gm/ccs, whether as an improved fallow, dispersed trees, green herbicides, grain legumes or fodder crops. Farmers may even be able to eliminate fallowing altogether, improved or otherwise.

Of course, there is a cost to keeping the animals out of fields: that of denying the animals access to crop residues. However, by using gm/ccs, farmers can often keep their animals better fed with a portion of the green gm/cc biomass than they can with dried out crop residues (see, for instance, S17). By maximizing biomass production, they can produce enough biomass to keep the soil fertile *and* feed more animals than they presently own.

Sometimes, the advantages of controlling the animals make it worthwhile for smallholder farmers to establish either social fences or physical ones to protect their gm/ccs, especially lablab beans or pigeon peas. Because both species produce most of their grain well after the dry season starts, they are largely useless when planted without fencing where domestic animals are free-grazing.

Probably the best way to deal with free-ranging cattle is to have one or two farmer leaders grow lablab beans for several years, either in fenced areas like family vegetable gardens or in other areas to which grazing animals have no access. Once villagers realize the value of the lablab bean for human consumption, marketing, dry season grazing, controlling weeds, fertilizing the soil and maintaining soil moisture, they will often make village-level decisions to limit free-grazing. This process of "social fencing" is greatly facilitated when the large-scale cattle-owners realize that with lablab beans in their own fields, they will have more than enough dry season fodder for their animals.

However, it is almost always difficult to get farmers to control their grazing animals when this is not already the custom. A program wishing to introduce the control of grazing animals should realize that it will take a good deal of time and dedication by program personnel.

If necessary, farmers can make cheap fencing by planting one of several different kinds of thorny plants around their plots. Such structures are frequently used around small vegetable or cassava gardens, but rarely around large fields. Probably the best species for this is the monkey thorn tree (*Acacia galpinii*), which makes a relatively thin fence of great strength that will last more than 40 years. This fence needs to be pruned each year to the desired height, but the trimmings also help fertilize the soil.

If people want to grow gm/ccs enough to establish fencing, social or otherwise, read on. If not, but they also consume cowpeas, rice beans, mung beans, peanuts or bambara groundnuts, go to the appropriate sections below.

Improved fallows present an excellent opportunity to use mucuna. Fallowing with perennial mucuna normally leaves the soil both rich in N and OM and close to weed-free within 12 months (photo 9). Farmers can broadcast or plant the mucuna seed any time there is little or no vegetation, the field has been plowed or cropped sometime within the last year, and it will rain fairly well for at least the next three to four months. Seeding rates vary from 2 to 4 seeds/sq m. If farmers live in a high-rainfall area and wish to use mucuna as an improved fallow, go to S52. If farmers would prefer to use tephrosia as an improved fallow, see S8. If farmers grow mucuna and also raise pigs, they can experiment with feeding mucuna to their pigs (see S54).

ICRAF, the World Agroforestry Center, recommends planting trees as improved fallows.[88] This approach requires much more labor than just broadcasting mucuna or jack beans, and several years will pass before farmers see the desired impact on fertility, moisture content or weed control. If the area's natural fallows last four years or longer and firewood can be sold at a good price, a tree fallow might be worth considering. In the vast majority of cases, however, a food-producing legume, or even one that produces no food but increases soil fertility within a year or two, would be preferable. That said, where rainfall is less than 400 mm/year, FMNR would be best.

Noxious Weeds

6. Are parasites and noxious weeds a serious problem? If your area is infested with striga (*Striga* spp.), imperata grass (*Imperata cylindrica*, called speargrass in Africa), nut grass (*Cyperus rotundus*) or wild sugarcane (*Saccharum spontaneum*), read on. If not, move on to the "Food Preferences" section below.

7. Farmers have serious problems with parasites or weeds. Striga grows only in soils that have been depleted of SOM. If the striga problem is reducing crop yields by less than 50%, farmers should probably let the gm/ccs solve the problem; the striga will naturally disappear over several years as gm/ccs increase the SOM content. However, if the yield reduction caused by striga is worse than 50%, the push-pull system could provide significant benefits for farmers. This system consists of intercropping greenleaf desmodium with the affected crops, and then planting hedgerows of Brachiaria or dwarf Napier grass about every 20 m across the field. The desmodium will "push" the striga away from the crops by inhibiting its growth, while the grasses will "pull" insect pests away from the crops.

Imperata grass also disappears gradually as the SOM content increases, but in the meantime the damage to crops is often much greater than with striga and the disappearance of the imperata grass is slower. Therefore, it is nearly always worthwhile to make a special effort to eliminate this weed right from the start.

If the problem is imperata grass, see S11 and S12. If it is nut grass, see S13, and if it is wild sugarcane, see S14.

S11. Control of imperata grass/mucuna. Proven. Benin. Roughly 10,000. This system, introduced by the International Institute of Tropical Agriculture (IITA), uses mucuna to control imperata grass (*Imperata cylindrica*). Imperata grass is one of the world's most noxious weeds, yet mucuna can turn imperata-dominated wastelands into cultivated fields.

The specific techniques vary somewhat, depending on the level of imperata grass infestation. Farmers usually burn the imperata grass and then plant mucuna. Then they probably will have to chop down the imperata grass again to let the mucuna develop sufficiently to smother it. The imperata grass will die after being completely shaded by the mucuna for four to five months. Often farmers must then eliminate the last 5 to 10% of the imperata grass by hand. Figures in the literature vary, but most agree that more than 10,000 farmers have used this system.

Leucaena and gliricidia trees can also control imperata grass, and do so with much less effort, but they take 4 to 5 years to accomplish it.

S12. Control of imperata grass/tithonia. Proven. Philippines. Thousands. Farmers traditionally slash the imperata grass on wastelands dominated by imperata grass and then plant meter-long cuttings of tithonia in contour rows 2 to 3 m apart. The tithonia largely eliminates the imperata grass in one year, but farmers leave the tithonia growing for a second year to fertilize the soil. The third year, the land is fertile and weed-free, ready to be cultivated.

S13. Control of nut grass/mucuna. Proven. Vietnam and Honduras. Thousands in each country. Nut grass (*Cyperus rotundus*), one of the world's most noxious weeds, is almost impossible to eliminate from a heavily infested field. However, farmers in parts of both Vietnam and Honduras eliminate nut grass by planting mucuna densely in an infested field and allowing it to completely cover the nut grass for at least six months, thereby killing the nut grass.

S14. Control of wild sugarcane/mucuna. Proven. Panama. Hundreds. Wild sugarcane was imported into Panama in the early 20th century to stabilize the steep earthen banks along the Canal. It then became a noxious, invasive weed. Farmers use mucuna to control it, in a way similar to its use in S11.

Food Preferences

8. Do farmers consume or grow any grain legumes? If farmers grow or consume any grain legumes, read on. If not, or if farmers are already using these legumes well as gm/ccs, go to the "Dominant Crops" section below.

What grain legumes do local people know or consume? Which ones command a good price in local markets, with plenty of demand? If they are lablab beans or pigeon peas, read on. If they are cowpeas (*Vigna unguiculata*), rice beans, mung beans (*V. radiata*), peanuts or bambara groundnuts, go to the appropriate sections below.

As already mentioned, the edible legumes mentioned in this section may work quite well as gm/ccs in bimodal areas, because the loss of N and biomass will not be nearly as great during only a 3- or 4-month dry season.

9. Lablab beans. Use any of the four systems below:

S15. 1st maize/lablab bean. Proven. Honduras, Peru, Ethiopia, Kenya, Rwanda, Tanzania, Malawi and Mozambique. Tens of thousands. Farmers near Guinope, Honduras, intercrop lablab beans with their maize. They plant both crops at the same time in May or June when the rains start, using two to four lablab seeds/sq m (photo 13). In this case, farmers are motivated largely because the lablab beans control the nematodes that used to damage their irrigated dry season garlic crop. About 75 farmers use this system, which Elías Zelaya, a farmer experimenter, developed 32 years ago.

Tens of thousands of farmers use a similar traditional system in northern Peru between Cajamarca and San Ignacio, where the lablab beans are traditionally eaten as green peas. In Kenya, tens of thousands of farmers produce black-seeded lablab beans, while in Tanzania thousands of farmers produce black- or brown-seeded lablab beans in a similar system. In both cases they are eaten or sold as dry beans. Lablab beans are sold as "black beans" in even the most exclusive of Kenya's and Rwanda's supermarkets. In large parts of Ethiopia, Malawi and Mozambique, farmers also grow lablab beans this way for local consumption as dry beans.

S16. 2nd maize/lablab bean, cowpea and rice bean. Proven. Thailand. Two or three hundred. Farmers relay their lablab beans, rice beans and cowpeas into their maize, as a 3-year rotation with one legume each year (see S36 and photo 45).

S17. 3rd maize/lablab bean. Good Potential. Honduras. At least 50 large-scale ranchers. During the last 25 years, large-scale cattle ranchers in southern Honduras have begun intercropping lablab beans with their maize in fenced-in fields. They plant the lablab at the same time as, or up to 2 weeks after, the maize. Once the maize and lablab beans have been harvested, their cattle have a plentiful, green, very palatable and highly nutritious fodder throughout the 6-month dry season. The ranchers do not need to cut the maize stalks or cut and carry the lablab fodder; they just close the gates to the field all dry season long, so the cattle can graze the maize stalks and green lablab biomass right in the field. (Note that in most of the lowland tropics, a number of gm/cc systems can produce large quantities of green fodder year-round. Making hay or silage would merely produce a less nutritious fodder at a much greater cost.)

Though this system is presently used only by large-scale ranchers, it should also work perfectly well for smallholder farmers with only one or two animals.

S18. Roof agriculture/lablab bean. Proven. Bangladesh. Tens, if not hundreds, of thousands. Land-hungry Bengalis often plant a few seeds of lablab beans (known as "hyacinth beans") right next to their houses, so the bean will grow up and cover their thatch roofs. They eat the seeds as a dry bean (or "pulse") and spread the greenery across their fields as fertilizer. Farmers would probably only accept this system where they have less than about 0.2 ha of land.

10. Pigeon peas. Farmers can use pigeon peas in several ways. If they want to use them to provide dispersed shade for vegetables or other low-stature crops, see S6. If they want to intercrop them with maize, see S19 and S20. If they might be interested in intercropping them with maize and peanuts or maize and sorghum as a "doubled-up legume," see S21. Or if they would like to plant them with upland rice, see S22.

S19. Maize/pigeon pea. Proven. Panama, Brazil, Zambia, Kenya and Malawi. Tens of thousands in the first three countries. Pigeon peas ("guandú") are widely consumed in Panama. Farmers intercrop them with maize as a gm/cc and a food source, planting about 1 seed/sq m of pigeon pea at the same time as the maize. Thousands of farmers east of the town of Machakos in Kenya also plant pigeon peas in their maize, but the pigeon pea population varies widely, often in the same field, with an average of only about 1 plant every 10 sq m.

S20. Maize/ratooned pigeon pea. Proven. Zambia and Malawi. Hundreds in Zambia, thousands in Malawi. In a slight but very valuable modification of S19, farmers intercrop pigeon pea with maize, planting both crops at the same time, with 1 or 2 seeds of pigeon pea/sq m. After the pigeon pea harvest (usually well into the dry season), they leave the plants standing in the field. When the rains come again, the farmers plant the new crop of maize in among the still standing pigeon pea plants. Ten days later, they cut off, or "ratoon," the pigeon pea plants at a height of 40 to 50 cm above the ground. Another 20 days later, they strip off the flush of new leaves to fertilize the maize.

Using this system, Sebastian Scott, near Lusaka, Zambia, has raised his maize harvests from 1 t/ha to 4 t/ha in just 6 years! This dramatic improvement in yields obviously occurred because of the increased biomass and N produced by the pigeon pea and the greatly improved synchronicity of the biomass application. The ratooned pigeon pea plants have also proven to be extremely drought-resistant, making them very popular in areas with rainfall down to about 300 mm/year. As an added benefit, ratooning saves farmers the cost of seed.

Sebastian also allows his cattle to graze some of the pigeon pea toward the end of the dry season. In this case, he plants these pigeon peas at 5 to 6 seeds/sq m because then the stalks do not become thick and woody, so they do not break off while the cattle are grazing.

To minimize the risk of losing all their pigeon peas during a drought, farmers should plant one quarter of their pigeon peas from seed each year, so when a drought does occur, they will rarely ever lose more than the one-quarter of their crop that is growing from seed that particular year.

S21. Maize/pigeon pea and peanut or soybean. Good potential. Malawi and Zambia. Thousands. This "doubled-up legume" system has been heavily researched, and farmer adoption is now beginning to take off. Of course, there are many gm/cc systems that use many more than two legumes in the same field, but this system has a lot of potential. One of its advantages seems to be that the peanuts grow extremely well because they are receiving dispersed shade from the pigeon pea.

S22. Upland rice/pigeon pea. Proven. Laos and Thailand. Thousands in each country. Farmers plant pigeon pea (*Cajanus cajan*) at the same time as upland (i.e. non-flooded) rice, with one pigeon pea seed every 1.5 to 2 m in each direction. The positive impact of this system may result from the pigeon pea's impact as a dispersed shade, as well as the nitrogen fixation and added biomass production.

11. Cowpeas, rice beans and mung beans. Cowpeas, rice beans and mung beans, all being from the genus *Vigna*, act in similar ways, except that some varieties crawl, some climb, and others are bushy. Bushy and crawling *Vigna* varieties can be intercropped with maize, sorghum, millet or cassava, but the climbing varieties should only be intercropped with maize, except on an experimental basis, because sorghum and millet stalks may not support the legumes' weight. The *Vignas* fix about 80 to 120 kg of N/ha as opposed to about 30 to 40 kg N/ha by common beans. Just switching from common beans to a *Vigna* species can make a big difference in soil fertility (assuming most of the N is not lost through volatilization).

In general, among the *Vignas*, rice beans tend to be more drought-resistant than mung beans. For the driest situations, or use as a relay crop toward the end of the wet season, short-cycle 60-day cowpeas may be the best option. In Vietnam, farmers commonly plant rice beans after paddy rice because the rice beans can produce a good yield using only the residual moisture in the paddies. Determinate 60-day cowpeas might well work in similar circumstances.

If you wish to work with cowpeas, you can choose from S23 through S31.

S23. 1st maize/cowpea. Proven. Honduras, Guatemala, El Salvador, Nicaragua, Panama, and most of Africa. At the very least, thousands in the first three countries, undoubtedly millions in Africa. Farmers plant cowpea (*Vigna unguiculata*) varieties intercropped with maize. They usually plant the cowpeas at the same time as the maize and within the maize row, to facilitate weeding. Cowpeas are locally known in southern Honduras as "alacín" and "pochote," and on the south coast of Guatemala as "frijol rienda" or "frijol de tierra." Under dispersed shade, the cowpeas can improve the soil.

S24. Maize and pigeon pea with cowpea or rice bean. Experimental. Zambia. None. The "doubled-up legume" system described in S21 could easily use cowpeas or rice bean in place of the peanuts. The pigeon pea shade would make good gm/ccs of the cowpea and rice bean.

S25. Various crops/cowpea. Proven. Cambodia. Several thousand. In southern Cambodia, farmers relay cowpeas into rice or vegetables, or plant them after these crops, to grow during the dry season, much like in S43.

S26. Rice/cowpea. Proven. Vietnam and Indonesia. Tens of thousands in each country. Farmers use cowpeas in the same way they use rice beans in S37. A similar system used to be traditional in Indonesia, but largely died out when the national government began subsidizing synthetic fertilizers. When the subsidies were ended many years later, however, tens of thousands of farmers abandoned fertilizer and revived their traditional gm/cc systems.

The crawling cowpea could very likely be profitably introduced to many new countries. Most easily, it could be used in fields that already have species of intercropped legumes that allow a fair amount of sunlight through to the soil surface, such as pigeon peas.

Crawling cowpeas can reduce maize yields if they are planted at the same time as the maize, so farmers should plant them one to three weeks after the maize. One should also note that some crawling cowpea varieties produce somewhat less grain than the bushy types.

S27. Upland rice/crawling cowpea. Proven. Laos. Thousands. Crawling cowpeas are planted sparsely among the rice. Some famers mix cowpea seeds with rice seeds at a rate of only 1 to 200 before broadcasting the mixture.

S28. Low-stature crops/crawling cowpea. Proven. Mozambique and Malawi. Hundreds of thousands. Certain crawling cowpeas do not climb at all. These include the normal crawling cowpeas that are presently grown widely in Mozambique, Malawi and some neighboring countries. Another is the 9-month Lojy Be cowpea, that, as far as we know, is only found in southwest Madagascar. These crawling cowpeas very likely have tremendous potential for any developing nations where cowpeas are presently grown and eaten. They can probably be intercropped with any number of low-growing crops, such as upland rice and medium-height vegetables.

Thus, we have varieties of cowpea that produce high-protein food, can be counted on doing so in even quite drought-prone conditions, are already very widely eaten virtually everywhere in Africa and tropical Asia, and in many cases will add an entirely new crop to farmers' fields above and beyond what they are already producing. If the crawling cowpea is Lojy Be, it will also have a significant impact on the long-term OM and N content of the soil.

The main limitation of any of the crawling cowpeas is that farmers can only manage them easily in fields where zero tillage, in-row tillage or hand hoes are being used.

S29. Most crops/Lojy Be Cowpea. Proven. Madagascar. Hundreds of thousands, with several thousand already using the drought-proof system. Farmers in southwest Madagascar plant the Lojy Be cowpea in January, a month or two after planting their maize or cassava. The spacing varies, but is approximately one seed/sq m. Within 45 days, the Lojy Be has covered the soil completely. It produces grain for four months, from April through July, and then stays alive, keeping the soil covered for another 3 months, until October. This means that of all the widely-consumed annual legumes, it is unique in being able to supply significant quantities of N to the next crop in unshaded lowland, unimodal areas.

Interestingly, farmers have two ways of managing the Lojy Be. Most farmers cut down the plant after the harvest and the residues are either eaten by cattle or lie on the ground through the dry season, doing nothing for the next crop. Yet, quite strikingly, in a 40- to 50-village area near Befandriana Sud, farmers leave the plant to grow until the following October, just before the next season's crops are planted. In a droughty year, this area has 2-m tall, dark green maize plants that produce about 3 t/ha of maize every year, while all around them, where the Lojy Be is cut down after the harvest, farmers are producing less than 0.5 t/ha (see photos 3 and 4). The Befandriana Sud system is spreading quite rapidly.

60-day cowpeas present us with another whole series of opportunities (see S30 and S31).

S30. Various crops/60-day cowpea. Proven. Mali, Burkina Faso and Mozambique. Thousands. Farmers can plant 60-day determinate cowpeas in most maize, sorghum, millet or cassava fields—even ones already intercropped—without affecting the growth of the other crops. This possibility exists because the 60-day cowpea will have been harvested and cut down before the other crops need the space occupied by the cowpea.

The 60-day cowpea can also produce a harvest if it is planted after a good rain and never receives any additional rain. That is, if the soil is wet when the cowpea is planted, farmers can be assured of a high-protein crop even during the worst of droughts. If, for instance, a good, heavy rain falls a month before the rainy season has started, farmers can sow these cowpeas and guarantee a harvest up to two months before any other grain. In this scenario, the cowpea can be incorporated into the soil in time to serve as a side-dressing of N for the main crop.

Thus, we have a crop that can be counted on to produce a high-protein grain even in extremely droughty conditions, may improve harvests and the OM content of the soil the first year it is used, will add an entirely new crop to most farmers' fields beyond what they already produce, and will cut short the hunger season by anywhere from one to two months!

The 60-day varieties definitely deserve to be tried out in drought-prone areas across the developing world.

S31. Most crops/60-day cowpea. Experimental. Malawi. A very few. Since the 60-day cowpea can produce a guaranteed harvest if planted immediately after a good rain, another possibility is opened up if there are good rains, or even just one good heavy rain,toward the end of the growing season. Planted immediately after this rain, the cowpea will produce a harvest on nothing more than the receding moisture.

Sometimes, however, insect pests could be a problem because of the lack of other biomass in the fields at this time.

If you are working with mung beans, select S32, S33 or S34.

S32. Cassava/mung bean. Proven. Mozambique. Tens of thousands. Farmers intercrop mung beans (*Vigna radiata*, known in India as "green gram" and in parts of Africa and Asia as "green bean") in their cassava fields. They tend to plant the cassava as far apart as 1.5 m X 1.5 m, and then plant several species of plants like mung beans, cowpeas, peanuts and/or sweet potatoes between the cassava plants whenever and wherever there is enough space and according to food needs and market prices.

S33. Rice/mung bean. Proven. Vietnam and Indonesia. Many thousands of farmers in each country. In Vietnam, farmers plant the mung bean after paddy rice, just as they do with rice beans in S37. A similar system has also been used traditionally in much of Indonesia.

S34. Maize/mung bean. Proven. Vietnam. Tens of thousands. In northern Vietnam, farmers traditionally intercrop mung beans in maize fields, planting both crops at the same time.

If you are working with rice beans, see S35, 36 and 37.

S35. Maize/rice bean. Proven. Mexico, Guatemala, El Salvador, Kenya, Uganda and Vietnam. Altogether, well over a million farmers. Farmers intercrop rice bean (*Vigna umbellata*) with maize, planting both at the same time. The rice bean is planted at a rate of around three seeds/sq m. Thousands of farmers use this system on the Isthmus of Tehuantepec in Mexico, tens of thousands on Guatemala's south coast and hundreds of thousands in El Salvador. Mechanization and counterproductive recommendations favoring monocropping have dramatically reduced the use of the system in El Salvador. A very similar maize/rice bean system is also practiced extensively in Vietnam and in small parts of Kenya and eastern Uganda.

People like the taste of the rice bean, but Latin American women find the tiny bean hard to thresh and clean. Anyone promoting the production of rice beans in Latin America should also teach the women how to winnow the grain as Asian women winnow rice.

Photo 45. Thailand. In northern Thailand, farmers rotate lablab beans, cowpeas and rice beans during the dry season following their wet season maize (see S36). As we add more gm/cc species to our systems, we may increasingly want to rotate the gm/ccs in order to keep pest populations under control.

S36. Maize/rice bean, cowpea and lablab beans. Proven. Thailand. Two or three hundred. In this high-rainfall environment, farmers have learned to relay rice beans the first year, cowpeas the second year and lablab beans the third year into maize fields, each one a month or two before the maize harvest. The legumes then continue to grow into the dry season. Farmers grow all these three legumes for multiple reasons: to fertilize the soil, to rotate their legumes over a 3-year period, to produce the legumes as food crops and cash crops, to suppress weeds during the dry season, and to keep the soil covered (see photo 45).

S37. Rice/rice bean. Proven. Vietnam. Hundreds of thousands of farmers, perhaps millions. Farmers traditionally plant rice beans immediately after the paddy rice harvest and at the beginning of seasons when the rainfall or water supply is insufficient for more paddy rice. This practice is so common that farmers in Vietnam believe the rice bean earned its name because it is for growing after rice, whereas most farmers around the world assume it is called "rice bean" because of the shape and diminutive size of the grain.

12. Peanuts and bambara groundnuts. Farmers often plant peanuts and bambara groundnuts together with and at the same time as, maize, sorghum, millet or cassava. Systems using peanuts or bambara groundnuts rarely fertilize the soil well, so farmers rarely plant them for that purpose. They are therefore not gm/cc systems by our definition. Nevertheless, where free-grazing animals are absent and dispersed shade or bimodal rains exist, farmers can use these crops to effectively fertilize and cover their soil.

S38. Maize, sorghum, millet, and cassava/peanut. Proven. Paraguay, Sahelian countries and Mozambique. Millions. Farmers very commonly intercrop peanuts with maize, sorghum, millet or cassava, to provide a cash crop, as well as a popular snack food and source of peanut sauce.

S39. Maize, sorghum, millet, and cassava/bambara groundnuts. Proven. Mali and several other Sahelian countries. Hundreds of thousands. Especially in seriously drought-prone areas, bambara groundnuts can be substituted for peanuts in all the systems using peanuts, with similar results as in S38.

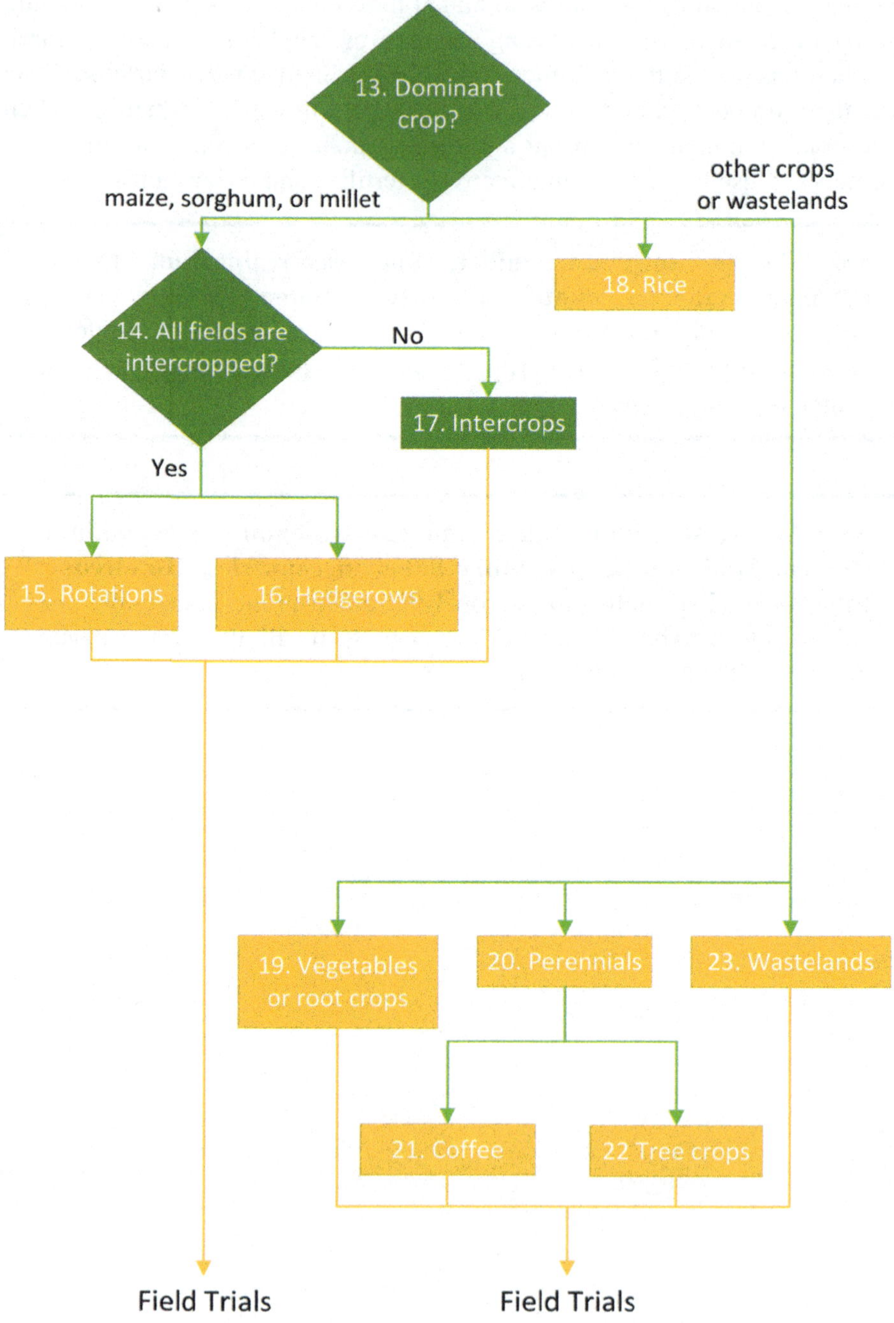

Numbers correspond to numbered commentary sections in Chapter 11

Chapter 11. Choosing Lowland Gm/cc Systems By Dominant Crops

13. Dominant crop. What is the major subsistence crop in the local farming system?

a) If the main subsistence crops are maize, sorghum or millet, continue on.
b) If it is rice, go to section 18.
c) If it is a vegetable or a root crop, go to section 19.
d) If it is a perennial, go to 20.
e) If a lot of the area's land is extremely dry or has become wasteland, go to 23.

Maize, Sorghum, or Millet Systems

14. Maize, sorghum or millet. Are many of the maize, sorghum or millet fields free of intercropped species? If so, go to section 17. If not, continue here.

Are all fields intercropped? First, make sure this is really the case. Many farmers feel they should intercrop all their fields, and therefore claim that they do. One should therefore personally observe a good number of the fields in the area to confirm how many are actually intercropped. If a large portion of the fields turns out to be monocropped, go to section 17.

Where almost all the fields are intercropped, the best approach is usually to plant dispersed shade, plant gm/ccs in a hedgerow or (if the rains permit) use gm/ccs in a dry-season rotation. The last two systems are usually less popular than other gm/cc systems, but they have been successful and sustainable in many countries. If farmers would prefer a dispersed tree system, see S3. If they would prefer a dry season rotation, continue here. If they prefer hedgerows, go to section 16. If they are not interested in any of these, and enough light is reaching the ground through the existing intercrop, try adding a crawling cowpea like the Lojy Be (S28) or a 60-day cowpea (S28 and S29).

15. Relays and dry season rotations. In order to take advantage of the dry season for growing legumes, farmers often relay grain legumes into the main crop (assuming the early season intercrops are harvested before the main crops) or plant the legumes after the main crops, at the beginning of the dry season (photo 45). In each case, the grain legumes grow mostly during the dry season. Other rotations use a whole growing season, which is less desirable, because this occupies rainy season land that could be used otherwise. The limiting factors for these dry season rotations include the length of the wet season, rates of soil moisture retention and whether or not any rain falls during the "dry" season. Dry season rotations are usually, though not always, used in areas where the total annual rainfall is over 1,200 mm.

For relays or dry season rotations, you can choose any system from S40 to S44.

S40. 1st maize/mucuna. Proven. Guatemala, Nicaragua, Paraguay, Brazil and Malawi. Only a few hundred in Guatemala but thousands in the other countries. In the Cerro San Gil region of Guatemala, farmers relay mucuna into maize 30 to 40 days before the maize harvest. Paraguayan farmers do the same 40 to 60 days before the maize harvest. Planted this late, the mucuna will often keep growing through much of the dry season, rather than dying in June (typical for the southern hemisphere) or in December (for the northern hemisphere).

S41. 2nd maize/mucuna. Proven. Honduras. A few hundred. In the 1970s, a few Hondurans tried to intensify the 3rd maize/mucuna system (S52) by growing two crops of maize a year. This complicates the management of the mucuna, but pays off in increased productivity. The farmers plant maize in both January and May. The mucuna has mostly died out in December, but drops seeds that sprout in January/February. By April, to prepare for the second maize crops, farmers cut down all the mucuna or eliminate it with an herbicide, and then replant it.

S42. Dry season rotation/tithonia. Experimental. Honduras. About 50. In this recently developed system, farmers around Sabana Grande in central Honduras sometimes broadcast tithonia seeds into their fields after the maize harvest in December. The tithonia grows throughout the dry season, providing added fertility for the following year's crops.

S43. 1st maize/jack bean. Proven. Cambodia. Thousands. Farmers traditionally use jack beans as a dry season relay or rotation with maize, in a manner similar to that of the cowpea in S25.

S44. Rotation/stylosanthes. Fallback. Philippines. Hundreds, but it was still being heavily promoted. On Leyte Island, a few farmers use stylosanthes species in rotation with several crops.

16. Hedgerows or alley cropping. Alley cropping is a kind of hedgerow system in which the hedgerows are normally only 6 to 8 m apart. Also, in alley cropping systems, the hedgerow is usually cut down close to ground level to prevent excess shading of the main crops. On the other hand, in hedgerow systems farmers may never prune the hedgerows or may cut them to almost any height. Sometimes they only cut the hedgerows when they need fodder.

Photo 46. Honduras. Tithonia is a "phosphorus accumulator," so farmers can use it as a gm/cc in areas where P is the limiting factor in crop production. However, such P deficiencies are rare.

Farmers do not tend to like hedgerow systems, especially alley cropping systems, because these systems occupy land needed for growing crops. Furthermore, they often require pruning during the growing season, when the demand for labor is highest. In fact, farmers have almost universally rejected alley cropping as it was originally designed (with 6-m-wide alleys), despite the huge sums of money spent promoting it.

Several modifications can help make hedgerows more popular. First, we should consult farmers about the distance between hedgerows. If they choose (as they likely will) a distance like 25 m between hedgerows, we should agree, even though the impact on soil fertility will be minimal because of the large distance. In time, when they see the impact of the hedgerows on soil fertility and soil moisture levels right near the hedgerows, they may decide to add an additional hedgerow half-way between each of the two they already have. Alternatively, they might prefer to maintain the hedgerows at 25 m and plant a double row of legumes in each hedgerow.

Second, if dispersed trees are planted in the hedgerows at the time the hedgerows are planted, the hedgerows will be less objectionable because they will be occupying the area of maximum shade, where crop growth might be a little less vigorous. Besides, the dispersed trees will provide added benefits, such as moisture retention and shade for crops. The trees will also usually provide enough added biomass so the system will at least maintain the soil's fertility, even if the rows are 10 to 12 m apart where crop growth might be a little less vigorous.

Third, on hillsides, the hedgerows can take on the added function of controlling soil erosion. On the island of Sumba in Indonesia, contour hedgerows of leucaena and gliricidia on hillsides have met with widespread acceptance.

S45. Hedgerow/tithonia. Good Potential. Honduras. About 50. A few Honduran farmers plant the tithonia as a contour hedgerow, cutting and spreading the branches across the field each year before the rains come.

S46. 1st hedgerow/tephrosia. Experimental. Honduras. Maybe 50. Farmers plant tephrosia in contour hedgerows and then cut and scatter the OM across the fields right before planting their staple crops.

Fourth, some hedgerow species, such as *Piliostigma*, can be pruned to ground level each year, which gives the associated crops a good head start. Since the *Piliostigma* grows fairly slowly, it requires no pruning during the rainy season—a tremendous advantage over the heavily promoted alley cropping systems.

S47. 2nd hedgerow/tephrosia. Proven. Vietnam, Haiti, Kenya, Indonesia and the Philippines. Tens of thousands. Vietnamese farmers, especially in Son La and Thai Nguyen Provinces, plant *Tephrosia vogelii* as contour hedgerows to recuperate eroded hillsides. Once the erosion is under control, the system becomes identical to S46.

S48. Alley cropping/gliricidia. Good potential. Zambia. Thousands. COMACO, a very innovative NGO in eastern Zambia, promotes an alley cropping system using gliricidia trees that are pruned as bushes. Farmers plant one seedling/m along the row with about 8 m between rows. Although some farmers would prefer the alleys to be wider (i.e. more maize plants and fewer rows of gliricidia), the system has definitely become popular. This popularity is not only because of the gliricidia's doubling of maize harvests from about 1.3 t/ha to around 3 t/ha, but because its contribution to SOM has brought a tremendous resistance to drought.

A very simple improvement on this system would involve pruning one gliricidia every 5 m along the row as a tree, rather than a bush. Such a system would provide dispersed shade, a lot more OM and even more drought-resistance.

The systems in S49 and S50 violate the gm/cc rule of not spending any effort on transport, but thousands of farmers use them in Burkina Faso. After all, they do usually require less labor than collecting and taking animal manure out to the fields, which tens of millions of farmers do every year.

S49. 1st sorghum/*Piliostigma.* Fallback. Burkina Faso. Many hundreds, if not thousands. In northeastern Burkina Faso, farmers traditionally collect leaves of the *Piliostigma* from wastelands and forests to fertilize their sorghum. I do not recommend this system because of the labor cost of transporting all that biomass, but the effort being made is a very good indication of the major impact this species' leaves have on the fertility of the soil.

S50. 1st millet/*Piliostigma*. Fallback. Burkina Faso. Thousands. Farmers use the same system as S49 to fertilize their millet.

We need to experiment a lot more with gm/cc systems for the Sahel, because farmers there have a serious lack of known gm/cc systems and they desperately need to improve both their soil fertility and their soil moisture retention. We now know that solving this problem is not impossible, because several gm/cc systems have already proven they can significantly improve soil moisture and fertility in such difficult environments: the Dogon system in Mali (S67), FMNR in the Sahel (S94), and the gliricidia alley cropping system (S3).

S51. Alley cropping/*Piliostigma*. Good Potential. Mali. A few hundred. The farmer plows a field with depleted soil and then plants *Piliostigma* seedlings, using either nursery seedlings or seedlings that volunteer near mature bushes. The seeding density will depend on the number of seedlings available, but they could be planted to form hedgerows 12 to 25 m apart, with 1 seedling/m in each row. With *Piliostigma*, soil improvement occurs more slowly than with jack beans or gliricidia, because *Piliostigma* grows more slowly. Nevertheless, *Piliostigma* has one tremendous advantage over jack beans: it will be permanent, rather than having to be replanted every couple of years.

Tithonia can be used as a hedgerow (S45), a dry season rotation (S42), a way of controlling imperata grass (S12), and, I should think, as an intercrop. We need to encourage farmers to experiment with tithonia, especially in areas with acid soils where a lack of available P may be limiting crop yields.

17. Intercrops. By and large, monocropped maize fields provide more potential for growing a variety of gm/ccs than almost any other situation except that of outright wastelands. Maize plants are tall, which means the gm/cc can be of almost any height; maize's long narrow leaves allow a lot of light through the canopy so the gm/cc can receive its share of sunlight; maize is generally grown in clumps instead of just one seed to each hole, which also lets through plenty of light; its tall, strong stem allows climbing gm/ccs to do well; and maize is generally grown in areas of more generous rainfall than are sorghum or millet, the other common subsistence crops with a similar architecture. These factors explain well why so many gm/cc systems involve maize.

Photo 47. Honduras. The 3rd maize/ mucuna system covers hundreds of thousands of hectares of hillsides in Mexico, Guatemala and Honduras. Except for a handful of false banana weeds, the three or four farms on this hillside are completely covered by mucuna.

In areas with more than 2,500 mm/year of rainfall in central Africa, the 3rd maize/mucuna system (S52) of Mexico and Central America might be tremendously useful. It definitely should be tried.

Mucuna can be used in maize systems in several ways, although farmers must always be aware of its aggressiveness. One way to avoid this problem is to use a bushy-type mucuna. Recently, an excellent farmer trainer named Kofi Boa introduced into Ghana a very good bushy-type of mucuna for just that purpose. Those working in Ghana or nearby could access it from Kofi (kboa55@yahoo.co.uk). A bushy type of mucuna also exists in South America, but it produces very little viable seed, so is of much less use to smallholder farmers.

S52. 3rd maize/mucuna. Proven. Mexico, Guatemala, Belize and Honduras. An estimated 25,000. This system only works where it rains nearly all year long, so Mesoamerican farmers can only use it along the Caribbean coast of Mexico, Guatemala, southern Belize and Honduras (photo 47). The mucuna covers the fields for 10 months, then dies in December. In January, farmers injection-plant maize seeds through the dead mucuna, and by February or March, the mucuna has reseeded itself naturally. Thus, in general, all the farmers need to do is plant the maize and harvest it. They do not have to do any weeding, tillage, planting of mucuna or cutting it down (except in occasional small patches), and no fertilizing—ever. I suspect that this system enjoys the lowest cost/ha of any maize system in the world.

The results of this system are stunning. Walking through a field used this way for a decade or more feels like walking on a soft mattress (see photo 7). Even on a 35% slope, with no soil conservation work and rainfall of over 2,000 mm/year, there is no erosion because the topsoil, usually more than a foot thick, acts like a sponge. On nearby fields without mucuna, farmers can only grow maize for two to four years, and then must fallow the field or apply expensive synthetic fertilizer. By using mucuna, farmers can plant maize every year for decades, with productivity gradually increasing up to over 3 t/ha, or about 3.5 t/ha if a little urea is added.

Smallholder farmers apparently developed this system in Guatemala soon after the United Fruit Company introduced mucuna into the country in about the 1920s to feed the Company's mules. The system first spread along the Polochic River Valley and then into much of El Peten in northern Guatemala. Nevertheless, its use along the Polochic River has recently become rare because much of the land there has been bought up by large landowners to produce sugarcane for making biofuels. Currently, something over 5,000 Guatemalan farmers use this system.

The same system was introduced into Mexico, or developed independently, by smallholder farmers in Mexico soon after it started spreading across Guatemala. An estimated 10,000 to 15,000 farmers use it within Mexico alone. It is still expanding through spontaneous adoption by farmers in Tabasco State and northward from Veracruz toward the U.S. border.

The system has also spread north into Belize and east half-way across northern Honduras. Although by the 1970s the system was used by an estimated 10,000 farmers along Honduras' north coast, its use has recently diminished considerably. One reason for this disadoption is the accidental importation into Honduras of itchgrass (*Rottboelia cochinchinensis*), one of the world's most noxious weeds. Itchgrass will start taking over a maize/mucuna field any time the farmer allows a patch of the field to remain uncovered by mucuna. The more important reason for disadoption is that large-scale cattle ranchers have been buying up the smallholders' hillside land ever since the floods during 1998's Hurricane Mitch washed thousands of heads of cattle from the coastal lowlands out to sea.

The maize grown using this system produces on hillsides a lower yield/ha than do the adjoining large-scale conventional fields on flat land, but because of the low costs/ha, the profit margin/ha of the maize/mucuna system is about the same, while soil fertility is being increased.

The expansion of the maize/mucuna system in all four countries has been entirely by a process of farmers teaching other farmers; no outside program has been involved in any major way.

S53. 4th maize/mucuna. Proven. Ghana, Benin, Kenya and Mexico. Thousands in each country. This system simply consists of a normal intercrop of mucuna with maize. Farmers plant the mucuna late enough that the maize is not overcome by the mucuna. Doing so can often be tricky because of the mucuna's aggressiveness, but with enough experience can be successful.

Planting dates for the mucuna vary from 15 to 50 days after the maize is planted. The later dates are more applicable when the maize is short in stature, it is planted quite sparsely, farmers are reluctant to prune the mucuna or the area is drought-prone. (Mucuna is more resistant to drought than maize, and therefore will tend to grow better and smother the maize.) In Kenya, some farmers plant mucuna late, and then pull the lower leaves off the maize so the mucuna will still grow well. The later plantings basically become relay cropping systems, so there is no distinct line differentiating this system from S40.

S54. 5th maize/mucuna. Proven. Mexico. Hundreds, if not a thousand or more. The difference between this system and S53 is that farmers use the mucuna seeds to feed pigs. In Mexico, fattening pigs is a very lucrative business if one does not have to buy commercial feed. In the southern Yucatán Peninsula town of Xpujil, farmers intercrop mucuna with maize, and then feed the mucuna seeds—boiled, ground and mixed half and half with maize meal—to their pigs. Introduced in the 1990s, the system has been spreading spontaneously since then.

S55. 6th maize/mucuna. Proven. Honduras. 200 to 300. Around the town of Omoa on Honduras' northwest coast, where more than 2,000 mm of rain falls each year, farmers have developed yet another maize/mucuna system. They plant maize in May/June, intercropped with mucuna. After the harvest, they cut down the mucuna to form a dense mulch and injection plant maize seed through the mulch, with a dibble stick, in September, without planting mucuna a second time. The mucuna mulch allows them to grow the second maize crop without ever having to weed it (photo 48).

Photo 48. Honduras. This second-season crop of maize is tall enough that it would normally need a second weeding. Nevertheless, under the 6th maize/ mucuna system (S55), the mucuna mulch is so thick that this maize crop never needed any weeding. This case, along with many others, shows that agroecology often requires less labor than does chemical agriculture.

Gm/ccs are rarely found in irrigated fields because of the high opportunity cost of the land, but one such existing system, the 7th maize/mucuna system (S56), might give some idea of other possibilities.

If the soil is too infertile for mucuna to grow well, farmers can grow jack beans instead. Jack beans fix a lot more N (up to 240 kg/ha), but will usually not provide the same degree of weed control. The bushy-type jack bean can be grown alone or associated with maize, sorghum, millet or cassava, whereas the climbing jack bean should only be used with maize.

S56. 7th maize/mucuna. Proven. Mexico and Costa Rica. A few hundred. This system has existed for about 35 years along the Isthmus of Tehuantepec. It is used on irrigated land, with the mucuna intercropped among the maize 30 to 40 days after the maize is planted. Often the farmers must prune the mucuna to keep it from covering the maize, although this will depend on the height of the maize, how many months the maize requires to mature, etc. Different planting patterns are used; one is to plant three mucuna seeds per hill of maize, but only in every other row. The planting season depends on the supply of irrigation water.

Other possibilities of lowland gm/ccs to use with maize include:

S57. 2nd maize/jack bean. Proven. Mexico and Brazil. Thousands, if not tens of thousands, in each country. Farmers traditionally intercrop jack beans with maize in several areas of northern Yucatán and the state of Oaxaca. They plant bushy-type jack beans at the same time as the maize or up to 30 days afterword, with two or three seeds/sq m, or four if weeds are causing serious problems. In Brazil, both jack beans and *Canavalia paraguayensis* (with a tan-colored seed) are intercropped with maize.

S58. 3rd maize/jack bean. Experimental. Honduras. Just a few. In the past, farmers in the drought-prone area near the town of Sabana Grande could only grow one crop of their 90-day maize varieties during the first half of the very unreliable 5- to 6-month rainy season. Yet after three to four years of intercropping jack beans with their maize, the soil retains enough moisture that farmers can grow two crops of maize and jack bean each rainy season. Furthermore, the soil is much more fertile. One farmer who used to harvest 200 kg/ha/year of maize now harvests over 1.6 t/ha/year of maize between the two plantings.

S59. Maize/soybeans. Proven. China, Southeast Asia. Hundreds of thousands, if not millions. In areas where soybean consumption is traditional, farmers often intercrop soybeans with maize.

S60. Maize/leucaena. Fallback. Benin and Ghana. Thousands. In another interesting system, farmers intercrop leucaena trees with maize at a rate of about 3 seeds/sq m. After the maize harvest, farmers do a light burn to kill the young leucaena trees, because doing so is much easier than cutting their woody stems. They purposely use a light burn (i.e. they burn the field when it is still somewhat moist) in order to avoid burning away all the N. Use this system only as a last resort because burning should never be done unnecessarily.

S61. Maize/lima bean. Proven. Mexico, Guatemala, Honduras and Nicaragua. Tens of thousands in the Americas, perhaps millions in Africa. Lima beans, locally called "ibes" in Mexico, are intercropped with maize on the Yucatán Peninsula. Farmers plant the lima beans from the same day as the maize up to a month later. A similar system is used in western Guatemala, southern Honduras and Nicaragua, as well as across the humid tropics of Africa.

S62. Maize/sunn hemp. Proven. Tanzania. Tens of thousands. Sunn hemp is intercropped with maize, both to fertilize the soil and to provide a pesticide to control insects during grain storage. Farmers broadcast the sunn hemp seed at the same time as the maize, at a rate of about 30 kg/ha. They mix one part of sunn hemp seed with two parts sand to get a fairly even distribution of the seed. Tens of thousands of farmers used this subsidized system for years, but disadoption has been heavy since the subsidies were discontinued and fertilizer companies started pushing hard against it.

S63. Maize/crotalaria. Proven for large-scale farmers, experimental for smallholder farmers. Brazil. Thousands to tens of thousands. Farmers intercrop any one of at least three species of crotalaria (*Crotalaria grantiana*, *C. juncea* and *C. mucronata*) into maize fields in Santa Catarina State (photo 49). They generally plant the crotalarias at the same time as the maize. In Brazil, all these systems, including those using forage turnips, sunflowers and oats as gm/ccs, were designed for middle-sized to large-scale farmers. Farmers adopted them not so much to provide soil nutrients as to produce huge amounts of biomass so they could move to zero till systems within a year or two. Since 2000, however, they have found that they can adopt zero tillage the same time they start planting gm/ccs.

Photo 49. Brazil. These medium-scale farmers in Brazil use gm/ccs like this crotalaria to maximize biomass production and weed control. It has, in fact, produced so much biomass that it has completely hidden from sight the already-harvested maize with which it was intercropped.

Anywhere that farmers intercrop maize and millet, as in southern Honduras, system S64 might be useful.

S64. Maize and millet/cowpea. Good Potential. Honduras. Hundreds. In the drought-prone Department of Choluteca in southern Honduras, farmers intercrop maize and millet in the same fields. They do this to lower their risk in case of drought. If they have a good year, they can eliminate the millet and produce a good maize crop. However, if a drought hits, the maize will be stunted and most likely the leaves harvested for animal fodder, while the millet will usually produce a decent yield.

To also have a gm/cc, some farmers reduce their risk even more by planting cowpeas among the maize and millet. Since one of the two major crops is eliminated within the first month or two, the cowpea normally gets plenty of sunlight. The cowpea provides an extra high-protein food source, and fertilizes the soil well where farmers have established dispersed shade.

Millet and sorghum. For millet and sorghum, farmers can use any of the above maize systems, as long as the gm/cc does not climb. In the drier areas of Africa, farmers can use *Piliostigma* with millet or sorghum.

S65. 2nd millet/*Piliostigma*. Proven. Burkina Faso and Mali. Thousands. In this system, farmers allow *Piliostigma* bushes to volunteer in their fields. Just before the rains come each year, they prune the bushes down to ground level and spread the leaves across the field to fertilize it. This traditional practice, used by farmers in both northeastern Burkina Faso and in Mali, is most visible along the highway just southwest of Segou.

Another possibility for drier areas was discovered recently (see S66).

S66. Sorghum, millet, cassava/bushy lablab bean. Experimental. India and Tanzania. Unknown numbers. Neil Miller in Tanzania recently discovered a bushy-type variety of lablab beans in India that could likely be intercropped with sorghum, millet and cassava, in addition to the maize. If so, this very valuable gm/cc variety could and should be used in a tremendous number of new areas across the world.

In areas where grazing animals are not a problem, relay systems in which the legume grows well into the dry season can also be used. (See, for example, S24)

S67. Intermittent shade/many gm/ccs. Proven. Mali. Two or three thousand. This system, developed by the Dogon people northeast of Koro in Mali, has basically made a 15-village area of Mali almost totally resistant to drought—a veritable oasis in the middle of a desert—with no special source of water. To develop this system, farmers have used a traditional FMNR system to allow local trees to grow. Then they pruned the trees in a funnel shape, narrow toward the base but wider toward the top of the tree. Because most of the foliage is therefore high off the ground, each tree's shade moves across the field during the day. Even crops planted within 1 m of a tree trunk do not get too much shade. "Dispersed shade" thus becomes "intermittent shade." This system allows for more trees/ha with denser foliage than could be grown in a dispersed shade system, thereby producing more biomass without negatively affecting the crops beneath.

The Dogon people then supplement the OM and N from the trees by planting several edible gm/ccs under the trees. These include cowpeas, peanuts, bambara nuts and the edible grass *fonio*, either intercropped in their millet or in small monocropped plots. A beneficial symbiosis is achieved: the shade from the trees makes the other crops grow better and reduces the amount of N volatilized through the dry season, while the legumes improve the soil and thus help the trees grow faster. The combination of crops has clearly improved soil fertility and moisture retention over time, and the farmers' previous striga problem has completely disappeared.

Rice Systems

If you need gm/ccs for rice, continue. For gm/ccs for vegetables or root crops, go to "Vegetables and Root Crops." If you need gm/ccs for perennial crops, go to the section below on "Perennials." If you need gm/ccs for wastelands or extremely drought-prone cropland, go to "Extremely Drought-prone Croplands or Wastelands."

18. Rice. As a very short-stature crop, rice can be difficult to intercrop. Even so, we are gradually finding gm/cc systems for both paddy and upland rice.

For upland rice. Farmers in Laos and Thailand sometimes intercrop pigeon peas, which provide a low-level dispersed shade, or let crawling cowpeas spread under their rice. Choose S22 or S28.

For paddy rice. Paddy rice has been a particularly challenging situation for using gm/ccs, because of the need for the gm/cc to grow in standing water. Most of these systems are rotations or relays. The only real intercrop is S69, which is an exciting new possibility we just heard about very recently. If you need a gm/cc for paddy rice, see systems S26, S33, S37, S68, S69 or S70.

S68. Rice/sesbania. Proven. Many countries in South and Southeast Asia. Tens of thousands, though the number is apparently decreasing as farms become smaller. Farmers plant *Sesbania rostrata* in rice paddies at the beginning of the rainy season to fertilize them for the subsequent rice crop. The sesbania is incorporated into the soil after it has grown, preferably, for about 45 days. Then rice is planted, as is done in traditional "green manure" systems from temperate climates. This system works best in areas where sesbania can grow during a 50-day period of light rains insufficient for irrigating rice.

S69. Paddy rice/crotalaria. Proven. Indonesia. Tens of thousands. We first heard about this system three months before delivering this book to the publisher. We do not even know which of the almost 700 species of crotalaria is involved. But we have been searching for over 20 years for a gm/cc that could grow together with rice in flooded paddy fields. Apparently this plant does that and a whole lot more. Therefore, in spite of our lack of information about the system, the tremendous potential this gm/cc seems to hold for tens of millions of farmers obliges us to mention it.

This crotalaria species is a spongy-stalked aquatic legume that grows well in paddy fields, produces a generous amount of biomass, produces N-fixing nodules on its stems, allows the rice to grow perfectly well and provides enough N to maintain soil fertility over the long haul. It even produces a relatively dispersed shade. It also decreases the C/N ratio of the rice stubble enough so the stubble rots quickly, ending farmers' felt need to burn the stubble. On top of all that, its tender shoots and pods are both edible. As a result, farmers growing rice in water-logged bottomlands wherever they exist across the island of Kalimantan in Indonesia traditionally grow this crotalaria in their paddy fields.

Farmers mix the seeds of the crotalaria with rice seed and broadcast them together onto the field.

S70. Rice/azolla. Proven. India and Southeast Asia. Hundreds of thousands. Azolla is an alga that grows in rice paddy water and fixes up to 30 kg of N/ha. It can be spread from paddy to paddy in the irrigation water to fertilize the rice virtually for free.

Other Systems

19. Vegetables and root crops. Because of the low stature and constantly changing crops in a vegetable garden, we have not found very many gm/ccs that work there. In most cases, the best way to integrate gm/ccs into vegetable gardens is to use a dispersed shade, such as S6, or a crawling cowpea, including either S28 or S29. More research on gm/ccs for low-lying, variable crops is needed and could prove to be very important.

The four systems presented here are meant for whole fields of a short-stature commercial vegetable crop (S71 and S72) or gm/ccs that are planted along the fence-line of a vegetable garden (S73 and 74).

S71. Tomatoes/jack bean. Proven. Mexico. Just over one hundred. Farmers intercrop jack beans in the rows between tomatoes, usually planting both at the same time. When the jack beans grow enough to start shading the tomatoes, farmers prune them back to a height of about 40 cm, causing them to grow laterally rather than vertically. The jack beans thereby control most of the weeds and fix a good deal of N. This system was developed in Yucatán by smallholder farmers during the 1980s.

S72. Chili pepper/jack bean. Proven. Mexico. Just over one hundred. Farmers use jack beans in their chili pepper fields the same way as in the tomato fields described in S71.

S73. Fence-climber/lablab bean. Proven. Philippines and Malawi. Tens of thousands. Many Filipinos on several of the nation's islands plant lablab beans to climb up the fences of their kitchen gardens. They do this in part because they have a variety with a soft pod they can eat like a snow pea. They also use the bean plant residues to fertilize their gardens. In parts of both northern and southern Malawi, farmers use the same system, but they consume the grain and the leaves of the lablab rather than the pods.

S74. Fence-climber/lima bean. Proven. Malawi and Ethiopia. Thousands. Lima beans are often planted along the fences around vegetable gardens in parts of Malawi and Ethiopia.

Photo 50. Paraguay. Because of the jack beans growing between the rows of cassava, this cassava crop has only required one initial weeding, rather than the two to three weedings a cassava crop normally needs. (The leaves of grass in the foreground are growing on the edge the road.)

Most of the systems we have for root crops outside of vegetable gardens are for cassava. Whether these would work for other root crops like yams or yam beans still has to be tested.

S75. Cassava/jack bean. Proven. Paraguay, Honduras and Mozambique. Tens of thousands. Farmers in southern Paraguay traditionally intercrop bushy-type jack beans with cassava (photo 50). They plant the jack bean between the rows of cassava, either at the same time as the cassava or up to 30 days later. The jack bean controls the weeds completely after the first weeding and provides enough N and biomass that within six to seven years cassava yields have usually tripled. Small groups of farmers in Honduras and Mozambique have taken up the same system.

When introducing this system, you should warn the farmers that there may well be a decrease in the first cassava harvest of up to 15 or 20%, because of the competition with the jack bean. For this reason, it is especially important that farmers experiment with the system on a small scale the first year and then increase the size of the plots over 2 or 3 years.

S76. Cassava/jack bean and mung bean. Experimental. Nicaragua. Hundreds. In a clever variation on S75, a group of innovating farmers in Diriamba, Nicaragua, fit some mung beans into the cassava/jack bean system. At the beginning of the rainy season, they planted cassava and mung beans. Then, when the mung beans had been harvested, they planted the jack beans. In this way, they achieved the soil fertility improvement of the jack bean (which grows easily through the dry season if it has received a couple of months of rain), but produced an additional crop of high-protein mung beans, too.

S77. Cassava/rice bean. Proven. Vietnam. Thousands. Vietnamese farmers intercrop bushy-type varieties of rice bean with their cassava, planting both crops at the same time. If the rice bean starts to compete with the cassava, the rice bean plants are pruned down to about 75 cm in height, making them grow laterally.

Vegetables or root crops with rotations or relays.

S78. Rotation with onions/mucuna. Proven. Brazil. Thousands, if not tens of thousands. Farmers rotate mucuna with onions in a system introduced by EPAGRI in Santa Catarina State.

S79. Rotation/yam bean. Fallback. Vietnam. Thousands, if not tens of thousands. In southern Vietnam farmers traditionally use the yam bean (*Pachyrhizus erosus*) in rotation with other crops as a food, a gm/cc and the primary ingredient for a home-made insecticide.

S80. Rotation/soybean. Proven. Vietnam. Hundreds of thousands, if not millions. Throughout Vietnam, farmers traditionally rotate soybeans with other crops to improve soil fertility and produce food. In general around the world, farmers use this system almost exclusively where soybeans have a good market or people traditionally consume them. Some new varieties of soybean that produce more biomass might be more useful as gm/ccs.

20. Perennials. Where farmers grow perennials as a main crop, it is almost always more efficient to use perennial species of gm/ccs. Perennial gm/ccs save labor in several ways: farmers do not need to replant the gm/ccs every season; they avoid weeding the field every year while the new gm/cc is getting started; and the perennials usually provide better long-term weed control than do annuals.

If the main perennial being grown is coffee, read on. If it is coconuts, oil palm, cacao, or fruit trees, go to the following section below.

21. Gm/ccs for coffee. Leguminous fertilizer trees are often recommended for shading coffee, but the recommended trees normally grow rather slowly, leaving the young coffee bushes unshaded for their first three or four years. In Guatemala, many farmers plant *Tephrosia vogelii* as a temporary shade for the coffee.

S81. Coffee/tephrosia. Proven. Guatemala. Thousands of mostly large-scale farmers. Farmers plant *Tephrosia vogelii* at the same time as the coffee seedlings. The tephrosia shades the coffee for about four years while the permanent tree cover grows large enough to take over the job. This practice has become popular in the last 40 years among both smallholder farmers and large-scale plantation owners.

Some of the most common shade trees for coffee are the *Inga* spp.

S82. Coffee/*Inga*. Proven. Guatemala, Honduras, Costa Rica and beyond. Hundreds of thousands. *Inga* spp. are used as a dispersed shade for coffee to help fertilize the soil and improve both the environment and the quality of the coffee.

Many smallholder farmers would like to produce additional income from their coffee fields, as well as protect themselves against those years when coffee prices crash. Some of these farmers have discovered that using some variety of fruit trees (citrus, bananas, papayas, etc.) to shade their coffee, instead of using traditional shade trees like grevillea (*Grevillea* spp.) or *Inga*, can improve both their incomes and their diets. In the years when the coffee price is high, these farmers prune their fruit trees heavily--to about 50% shade--in order to maximize their coffee production. When the coffee price is low, they prune their fruit trees sparsely, or not at all. In these years, their more valuable fruit production thereby increases at the expense of their relatively unprofitable coffee.

However, if farmers use fruit trees for shade instead of a gm/cc, they need to plant something beneath the coffee plants to maintain their soil fertility. Low-lying perennial gm/ccs work well for this purpose.

S83. Coffee/perennial peanut. Proven. Costa Rica and Nicaragua. Between five hundred and several thousand. Farmers in Costa Rica use perennial peanuts in coffee fields, much as they are used in Honduras for oil palms (S85). Groups of farmers in Nicaragua have started using the same system.

If farmers cannot obtain perennial peanut cuttings, they will often settle for jack beans.

S84. Coffee/jack bean. Proven. Costa Rica and Nicaragua. Thousands. Farmers intercrop bushy-type jack beans among coffee plants, both to fix N and to control weeds. They plant the jack beans about 50 cm apart in each direction across the field.

22. Fruit and other tree crops. For tree crops, perennial peanuts are almost always preferable, with jack beans and cowpeas serving as possible alternatives. Mucuna would have to be reseeded every year and has to be cut down around the trees. It would only seem to be advisable where farmers have a problem with sun scorch (see S87).

S85. Tree crops/perennial peanut. Proven. Honduras and Costa Rica. Several thousand large-scale farmers. Along the north coast lowlands of Honduras and in Costa Rica, farmers plant perennial peanuts around new oil palm seedlings to eliminate the weeds and supply N to the trees. Occasionally, the perennial peanut is used the same way to produce palm hearts in Costa Rica. Smallholder farmers can use the same approach.

S86. Tree crops/jack bean. Proven. Paraguay and Honduras. Several hundred. Farmers also use jack beans for fruit trees and other plantation crops. Jack beans require less work than some other cover crops because they can grow and maintain their cover for two or three years. Bushy types do not climb up into the trees.

S87. Citrus trees/mucuna. Proven. Paraguay. At least one hundred. In Paraguay farmers use mucuna as a gm/cc for citrus. Though replanting mucuna every year requires more labor than using a perennial gm/cc, these farmers have developed an ingenious little trick: by letting the mucuna climb up over the trees and completely cover them a month or two before the harvest, they prevent the fruit's being damaged by sun scorch. Right before the harvest, the mucuna is cut off near the base of the tree. This improvement in the quality of the fruit means farmers can export it to Brazil, where it earns about double the domestic price.

S88. Oil palm/*Desmodium*. Proven. Honduras and Belize. A few hundred. Farmers use *Desmodium ovalifolium* as a cover crop in oil palm plantations. Others in Belize use *Desmodium* this way in situations where the soil pH has dropped to below 5.0.

S89. Banana/mucuna. Experimental. Rwanda. About 25. Rwandans' banana production has suffered because the increasingly frequent droughts are causing the soil to dry out during the dry seasons. Some farmers have solved the problem by using mucuna as a cover crop. They have also tried lablab beans as a combination cover crop and food source, but the lablab failed because it is not as tolerant of shade as is mucuna.

S90. Fruit trees/cowpea. Proven. Thailand and Uganda. Thousands. Smallholder farmers also grow cowpeas under various fruit trees in a number of countries, investing additional labor in return for the high-protein food produced. They could probably use the other *Vignas* in the same way.

Photo 51. Paraguay. Farmers often plant gm/ccs among fruit trees more for weed control than for soil fertility (S91). In this photo, a lupine is growing among citrus trees. The lupines in Brazil largely disappeared years ago because of anthracnosis, a problem we now know can largely be eliminated by rotating the gm/cc species.

S91. Fruit trees/lupines. Proven. Paraguay and Brazil. Thousands. Farmers use lupines in Paraguay and Brazil as gm/ccs among citrus trees. At lower elevations, the perennial peanut would probably be better.

23. Wasteland or extremely droughty cropland. If the land is wasteland, see S92, S93 and S95. If land is under production but receives less than 500 mm of annual rainfall, see S94.

S92. Wasteland recuperation/perennial mucuna. Proven. Vietnam. Thousands. In northern Vietnam, people sometimes use a perennial mucuna species to recuperate wastelands. They broadcast the seed at the beginning of the rainy season on a piece of wasteland, and then allow it to grow until the land is judged fertile and largely weed-free—usually within a year, but sometimes two.

S93. Wasteland recuperation/Indigophera. Fallback. Vietnam. Probably thousands. Farmers sometimes use *Indigophera* trees (*Indigophera* spp.), just as mucuna or jack beans are used in S92 and S95. The main difference between the systems is that the *Indigophera* takes several years to make the soil fertile—a distinct disadvantage.

Photo 52 by Tony Rinaudo. Niger. The 24 million hectares where farmers are using FMNR are helping to push back encroaching deserts, including the Sahara (S94). Notice that in this photo, most of the trees' foliage is sparse enough or high enough off the ground that the millet under them is growing just as well as the millet out in the direct sunlight.

S94. Other crops/trees (FMNR). Proven. Niger and 10 other African nations. About 5 million farmers on close to 24 million hectares (confirmed by satellite photos). Farmer-managed natural regeneration of trees (FMNR), a very simple way of turning semi-desert areas into virtual forests or dispersed shade systems, is easily one of agricultural development's greatest success stories (photo 52).

FMNR provides a multitude of benefits, including firewood production, production of various other forest products, improved soils and increased resistance to drought. Perhaps most importantly, the shade, soil moisture retention, higher levels of SOM and general environmental improvement have created a very healthy platform for all sorts of additional agricultural innovations. Furthermore, almost all semi-arid areas of the world have what Tony calls an "underground forest." For decades, farmers burned their fields and treated tree sprouts as weeds to be eliminated. As a result, the root systems of 50 trees/ha or more may be living under the soil with very little indication above ground that they exist.

FMNR consists of merely allowing this underground forest to become an above-ground, managed forest. The process is simple. Instead of making nurseries and planting out seedlings (a labor-intensive process with a very high failure rate in such difficult environments), farmers merely quit cutting and burning their underground forests. Then they select which trees they want, allow the desired number of shoots to grow, prune all but 3 to 5 branches from each one, and protect them from free-grazing animals for the first year or two. This pruning must be done as early in the rainy season as possible. Trees grown this way achieve close to a 100% survival rate even in terribly drought-prone conditions, and they often grow twice as fast as trees grown from seedlings because they start out with very well-developed root systems.

By pruning the trees sufficiently, farmers can grow crops under them. The new trees then function as a dispersed or intermittent shade for their crops.

S95. Wasteland recuperation/jack bean. Proven. Vietnam, Mexico, Honduras, Mozambique and many other countries. Many thousands. Vietnamese farmers traditionally grow jack beans on wastelands to recuperate the land for the cultivation of other crops, usually within just one year. A week or so before the beginning of the rainy season, they broadcast jack bean seeds on the land at a rate of 3 or 4 seeds/sq m. Where weeds are particularly aggressive, farmers may need to cut them back once, to allow the jack beans to grow vigorously. In such cases, planting the seeds (rather than broadcasting them) might help because planted seeds germinate about two weeks sooner than do broadcast seeds.

The jack bean virtually always grows well in even the poorest of soils, as long as the pH is above 4.5. If the jack bean fails to grow well, it might be necessary to use a small amount of animal manure or compost to get it started.

S96. Wasteland recuperation/pigeon pea. Experimental. Brazil and Zambia. A few dozen. Pigeon peas have also proven their ability to recuperate wastelands. During their first year of growth, they are a bit less resistant than jack beans to poor soils and to drought. However, the second year, if they have been ratooned, they are more drought-resistant than jack beans and will fix nearly as much nitrogen. Furthermore, they produce both food and fodder. In areas where people already consume pigeon peas, this would be a technology well worth trying.

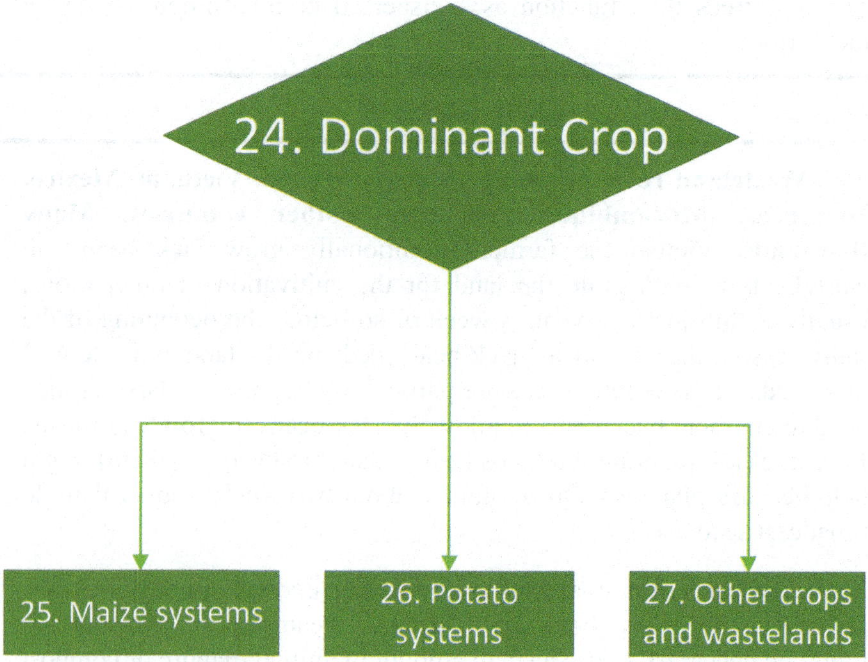

Numbers correspond to numbered commentary sections in Chapter 12

Chapter 12. Choosing Systems for the Highlands

Maize Systems

24. Dominant crop. At elevations above 1,500 m, there are fewer gm/cc species in use, but the number of dominant crops is also less varied. Most of these species grow well up to about 3,000 m, although tarwi does well up to 4,000 m.

25. Maize systems. If maize is the main crop in the area, read on. For potatoes, go to section 26 on potato systems. For any other situations, go to section 27.

S97. 1st maize/runner bean. Proven. Mexico, Guatemala, Honduras, Colombia, Ecuador, Peru, Bolivia and Chile. At least a million. Farmers intercrop runner beans with maize, planting up to one seed for every 2 sq m (photo 21). Farmers can maintain their maize yields for decades with this system. At the time of Columbus, it stretched through both temperate and highland tropical regions all the way from Canada through New York State in the U.S. (where it is known as the Seneca bean), through Mexico (*ayocote*), Guatemala (*piloy*) and Honduras (*chinapopo*) to Colombia (*frijol cacha*), Ecuador, Peru and Bolivia. It is still widely used in all the highland areas from central Mexico to Bolivia and northern Chile.

S98. 2nd maize/runner bean. Proven. Guatemala. At least 3,000. In the San Marcos Department of Guatemala, farmers intercrop runner beans with maize. The runner beans produce a tremendous amount of biomass—often more than mucuna. Nevertheless, since this variety of runner beans is a climber, the extreme weight of its biomass can cause even maize to fall over toward the end of the growing season. Therefore, farmers often plant the runner bean very sparsely, using only one or two seeds for each 10 sq m. They generally plant the runner bean simultaneously with the maize, and then allow its tuber to sprout the second and third years, making planting unnecessary.

The runner bean's production of a massive amount of biomass could provide a tremendous advantage in terms of soil fertility, moisture retention and maize productivity if the plant population could be increased. If the runner bean and maize could be supported by a third, fast-growing, woody-stemmed plant (maybe tephrosia), the system could become a very profitable highland maize-growing system. This possibility should be researched. Another, much simpler alternative, would be to import from Kenya the bushy-type runner bean, which apparently does not exist in Latin America (see S99).

S99. 1st maize/runner bean and common bean. Proven. Kenya. Hundreds, if not thousands. In the central highlands of Kenya around the town of Thika, farmers traditionally do something quite unique: they simultaneously intercrop two or three varieties of beans at the same time with their maize. Typically, they use one variety of white-seeded, bushy-type runner bean and two varieties of common beans (*Phaseolus vulgaris*), all of which grow to about the same height. I have only seen this bushy-type runner bean in Kenya and Zimbabwe.

The three beans are capable of more or less maintaining soil fertility over time. This surprisingly good impact on the soil for common beans is due in part to the cool highland weather, the presence of the runner beans (which fix about twice as much N as do common beans) and the bimodal rains, which only leave about three months between rainy seasons. The various beans, all together, control the weeds completely.

White-seeded runner beans have a very good market in both Kenya and Zimbabwe, in part because they are exported to the U.K., where they are eaten as green pods.

S100. Maize/bushy runner bean. Experimental. Mexico, Central America and the Andes. No one, yet. It is highly probable that the bushy-type, white-seeded runner bean from Kenya and Ethiopia could be used in many places in Latin America. This would allow the runner beans to be planted up to 2 plants/sq m, instead of just one or two plants every 10 sq m, as is now done in some places to prevent the maize from being pulled over. A denser planting would likely improve runner bean yields and increase OM and N levels in the soil. All we need is for someone to go through the legal phytosanitary process of importing the bushy runner beans from Africa.

S101. 2nd maize/runner bean and common bean. Proven. Honduras. Hundreds. Along the northern coast of Honduras, farmers intercrop maize with both runner beans and a variety of common bean called "frijol milpero," or "cornfield bean," which they say fertilizes the soil better than other *Phaseolus vulgaris* beans. In two-thirds of the planting holes, three seeds of maize and one seed of runner bean are thrown in the same hole. The remaining hole is planted with three seeds of maize, one runner bean seed, and one cornfield bean seed. In this high rainfall area, all three do very well.

S102. Maize/fava bean. Proven. Mexico, Guatemala, Peru, Ethiopia, Vietnam. Hundreds of thousands, maybe millions. Farmers grow fava beans intercropped with maize. They usually plant them about a month after the maize, at a rate of from two to five seeds/sq m.

S103. Maize/sweet clover. Proven. Mexico. About 200. This interesting system was developed by smallholder farmers in the 1950s. Designed for farmers who both grow maize and raise cattle, it is very popular in several villages near the town of Tlaxiaco in Oaxaca State. Farmers intercrop a perennial variety of sweet clover with maize, planting both at the same time. After the maize harvest, the sweet clover grows to about 2 m in height. The farmers' animals then graze down both the maize stalks and sweet clover until the end of the dry season. When the rains start again, farmers cut down to ground level what is left of the sweet clover and maize, then reseed the maize. The sweet clover sprouts again on its own. In this way, the sweet clover both maintains the soil's fertility for decades (assuming enough biomass is left by the animals) and provides fodder for the animals throughout the dry season.

S104. Maize/tailcup lupine. Proven. Ethiopia. Hundreds, if not thousands. Farmers relay this lupine into maize in areas above 1,500 m in elevation. As with tarwi, the tailcup lupine's grain can be consumed, but only after having been washed in a running stream for a couple of days.

S105. Maize/vetch. Fallback. Mexico. Over ten thousand. Farmers in Michoacán State intercrop vetch (*Vicia toluca*) with maize by planting both crops at the same time. The vetch serves to both fertilize the soil and feed animals during the dry season. After the maize harvest, donkeys line the highways, carrying piles of vetch back to people's homesteads. However, the sweet clover system (S103) would usually be preferable to this one, given that with sweet clover the animals can graze fresh clover all dry season long, without the need for farmers to cut, carry and store all the fodder.

Potato Systems

26. Potato systems. Farmers use several systems for growing tarwi to fertilize white potatoes in the Andean countries (see S106 to S108).

S106. 1st potato/tarwi. Proven. Peru, Ecuador and Bolivia. Hundreds of thousands. Farmers traditionally grow tarwi in a 3-year rotation with potatoes, both to fertilize the potatoes and to provide a high-protein grain. They plant tarwi one year, potatoes the next year, and a small grain the third year, followed again by tarwi. Over time, potato yields often triple if the field started out badly depleted.

S107. 2nd potato/tarwi. Proven. Peru and Ecuador. Tens of thousands. Farmers also traditionally grow tarwi in a single line around the edges of small patches of potatoes.

S108. 3rd potato/tarwi. Good potential. Peru. A few hundred. Farmers intercrop tarwi with their potatoes, doubling their previously low potato yields while also producing an edible grain

Other systems to use with potatoes include:

Photo 53 by Felipe Tomás Mux. Mexico. Fava beans produce a tasty grain that can be eaten green (like lima beans) or dry. Farmers in Guatemala and Mexico either intercrop or rotate them with basic grains.

S109. Potato/fava bean. Proven. Peru. Hundreds of thousands. Farmers traditionally plant fava beans in a simple rotation with potatoes.

S110. Potato/common pea. Proven. Peru. Hundreds of thousands. Farmers grow common peas in a rotation with potatoes just like with tarwi in S106.

Other Crops

27. Other highland possibilities.

S111. Rotation/fava bean. Proven. Mexico. Hundreds of thousands. Farmers grow fava beans in rotation with maize and other crops (see photo 53).

S112. Dry season rotation/chickpea. Proven. Ethiopia. Hundreds of thousands. Farmers plant chickpeas in a dry season rotation with many subsistence or cash crops across the Ethiopian highlands.

S113. Rotation/buckwheat. Fallback. Rwanda and Brazil. Tens of thousands. Farmers rotate buckwheat with other crops in parts of Rwanda and Brazil.

S114. Teff, wheat and barley/grasspea. Fallback. Ethiopia. Tens of thousands. Farmers plant grasspea as a traditional green manure during the light rains at the beginning of the season. When the rains become more reliable, they cut down the grasspea plants and plow them under, in time to plant wheat, barley or teff.

S115. Rotation/common pea. Proven. Brazil and Uganda. Tens of thousands. In the State of Santa Catarina in Brazil and in southwestern Uganda, farmers grow common peas in a rotation with a number of different crops. The peas serve as a food, a gm/cc and a market crop.

S116. Wasteland recuperation/tarwi. Experimental. Peru. Hundreds. Tarwi grows so well in poor soil that farmers can use it to recuperate wastelands (photo 54).

Photo 54. Peru. This wasteland in highland Peru, created when a road crew took away the top meter of the soil, is being restored using tarwi. In this case, the farmer used a small amount of manure so the tarwi would reach maximum growth. As a result, it will restore the soil enough to produce a good potato crop after just one year.

S117. Dispersed shade/alder. Proven. India, Myanmar and China. Hundreds of thousands. Farmers use alder trees (*Alnus nepalensis*) in dispersed shade systems all the way from Assam State in northeastern India to southern China and Myanmar. These trees are mostly planted at altitudes too high to help crops through their cooling effect, but the trees are easy to prune, provide the soil with OM, fix N and provide firewood and fodder for sheep and goats. As often happens when technologies spread so far, the technology changes from place to place. In China, for instance, most of the trees are not pollarded, but are replanted at the beginning of each fallow period.

ANNEX 1: ADDITIONAL READING

Major Sources

After the advent of major synthetic fertilizer use in the late 1940s, the English-speaking scientific community did not start researching gm/ccs seriously again for half a century, whereas Brazilian scientists began in the late 1970s. Brazilians have done extraordinary work on the subject but they publish their reports in Portuguese, and sometimes Spanish, in journals that few English speakers ever read.

For those who can read Spanish or Portuguese, I would heartily recommend the following three books, more or less in descending order of priority. The first two offer abundant, well-sourced evidence of much of what this book is saying:

1) Monegat, Claudino, *Plantas de Cobertura del Suelo: Características y Manejo en Pequeñas Propiedades* (Tegucigalpa, Honduras: CIDICCO, 1997). In Spanish and Portuguese. The original in Portuguese was printed in Chapeco, Santa Catarina, Brazil, and copyrighted in 1991.

2) Calegari, Ademir, et al., *Adubacao Verde no Sul do Brasil* (Rio de Janeiro: Assessoria e Servicos a Projetos em Agricultura Alternativa (AS-PTA), 1993). In Portuguese only. AS-PTA in Rio de Janeiro would probably be the best source.

3) Binder, Ulrike, *Manual de Leguminosas de Nicaragua, Volume I* (Esteli, Nicaragua: PASOLAC and the Escuela de Agricultura y Ganaderia de Esteli, 1997). In Spanish only.

I would also heartily recommend the one major book that has come out of the experience in Brazil and Paraguay that has been published in Spanish and English:

Florentín, M. A., *et al., Green Manure/Cover Crops and Crop Rotation in Conservation Agriculture on Small Farms* (Rome, Italy: Food and Agriculture Organization of the United Nations, 2011).

These books, aimed mostly at farmers who have 5 ha or more (which they call small-scale farmers, but in most of the developing world we would consider to be medium-size farmers), deal much more with non-edible gm/ccs than does *Restoring the Soil*, but it is nevertheless an invaluable book for those interested in gm/ccs.

Additional Books

Centro International de Divulgación sobre los Cultivos de Cobertura (CIDICCO), *et al., Experiencias sobre Abonos Verdes y Cultivos de Cobertura* (Tegucigalpa, Honduras: CIDICCO, 1998).

Eilittä, M., J. Mureithi, and R. Derpsch, *Green Manure/Cover Crop Systems of Smallholder Farmers: Experiences from Tropical and Subtropical Regions* (Dordrecht, The Netherlands: Kluwer Academic Publishers, 2004).

IFAD, *et al., Shifting Cultivation: Towards Sustainability and Resource Conservation in Asia* (Cavite, Philippines: International Institute of Rural Reconstruction, 2001). Pp. 238-243.

International Development Research Centre, Editors: D. Buckles, *et al., Cover Crops in West Africa/Plantes de couverture en Afrique de l'Ouest* (Sasakawa Global, 1999).

Kay, D., *Food Legumes* (Tropical Products Institute, 1979).

Mureithi, J., et al., eds., *Development and Up-Scaling of Green Manure Legume Technologies in Kenya* (Kenya Agricultural Research Institute and Legume Research Network Project: Nairobi, 2006).

National Research Council. *Tropical Legumes: Resources for the Future* (National Academy Press, Washington, D.C., 1979)

Rinaudo, T., *Farmer Managed Natural Regeneration: Exceptional Impact of a Novel Approach to Reforestation in Sub-Saharan Africa* (Technical Note #65, Educational Concerns for Hunger Organization (ECHO), 2012)

Website: http://edn.link/tn-home (ECHOcommunity.org)

ANNEX 2: GLOSSARY OF AGRICULTURAL TERMS

Allelopathic: Describes chemicals produced by one plant that make it difficult for other plants to grow nearby

Annuals: Plants that die within one year of germinating

Bimodal: Having two rainy seasons each year. In the tropics, bimodal climates tend to exist in a band along the equator.

Biodiversity: Refers to an environment containing many different species of plants, animals and microorganisms living together

Biomass: organic matter. Usually used when quantifying the amount of organic matter

Climate change: The gradual over-all increase of the world's temperature and other climatic changes that are being caused at least in part by people's use of fossil fuels, by the destruction of the world's forests and other natural environments and the decreasing amount of OM in forest trees and agricultural soils

Determinate: A plant that produces most of its fruit or grain at the same time

Dispersed shade: A condition of reduced sunlight in a field, produced by evenly spaced trees that produce fairly light shadows (usually reducing the sunlight reaching the soil by about 15 to 50%)

Exotic plant: A plant that is not native to a given region

Fallowing: The practice of temporarily leaving a field uncultivated for several years, to allow natural vegetation to grow back and to return the land to its natural state of fertility. Farmers traditionally used this practice throughout the tropical world, but because of population pressure on the land, it is now dying out.

Fixation: The biological process by which soil microorganisms take nitrogen from the air and turn it into a form that plants can use

Grain legumes: Leguminous plants that produce seeds that are eaten by humans

Green manure: A species of plant that is used almost exclusively to fertilize the soil. It is usually planted as a monocrop before a food or commercial crop, and is plowed under at flowering time. Farmers throughout the temperate countries used green manure systems before the 1940s, when synthetic fertilizers began spreading around the world.

Green manure/cover crop, or gm/cc: A species of plant, usually a legume, whether it is a tree, a bush, a vine, a crawling plant or an alga, which is planted by farmers to, among other things, maintain or improve their soil fertility or to control weeds. Farmers almost always have other reasons for growing these plants, too.

Hedgerow: Plants established in a line across a field to reduce erosion or to provide some other agricultural or environmental benefit

Inoculant: A commercial product containing microorganisms (plants or animals too small to see) used to increase nitrogen fixation on the roots or stems of certain legumes

Intercrop: A crop grown in the spaces between another crop or crops, planted within a month or two of the latter

Intermittent shade: Shade from tree leaves high enough above the ground that the shade moves as the sun does, thereby allowing the crops underneath to receive alternating sunlight and shade as the day progresses. The ultimate effect is the same as with dispersed shade, but allows for a greater total biomass/tree.

Invasive species: Plants that have gradually spread in areas where they were not planted or did not previously exist, and are damaging other productive activities or the environment

Legume: A plant with seeds that grow inside an elongated pod. These plants are particularly important in agriculture since, through the natural process of fixation, bacteria that live on their roots or stems take nitrogen out of the air and make it available to other plants.

Microorganisms: Animals or plants that are so small they can only be seen with the help of a microscope. A teaspoon of fertile soil can contain millions of them. Most of them are very beneficial to plant growth and health.

Monocrop: Planting only one crop in a field at a given time

Nitrogen (N): An essential plant nutrient that is the greatest limiting factor to soil productivity around the world. Nitrogen is becoming a more and more expensive part of synthetic fertilizer because of the increasing world price of energy.

Noxious Weeds: Those weeds that are harmful to plants, animals or the environment, and are particularly difficult to control

Organic matter: Anything that, in a previous form, was part of a living organism, such as parts of dead plants, bodies of dead animals, urine and manure

Perennial: A plant that naturally lives during two or more years

pH: A measure of acidity or alkalinity. A soils' pH will affect almost all plants' growth, especially if the soil is either extremely acidic or extremely alkaline.

Pollard: To prune most or all of the branches off a tree so new branches will grow out of the trunk

Ratoon: To cut a plant off at or near the soil surface, so that it will grow again using its existing root structure

Relay crop: A crop that is planted in a field where another crop is already growing, within two months before the initial crop's harvest

Rhizobium: A microorganism that often grows on the roots of legumes and fixes nitrogen. It takes nitrogen out of the air and puts it in the soil in a form that plants can use.

Rotation: A cropping system in which one crop is followed by another, or several others, in an established sequence. The sequence is usually designed to maintain soil fertility and reduce the weeds, pests, parasites, and diseases that affect crops.

Synchronicity or **Synchronization:** The degree to which nutrients or biomass are applied to the soil at the time when crops most need them

Transpiration: The process by which water evaporates from the leaves of growing plants

Unimodal: Refers to a climate with only one rainy season each year

Volatilization: The process by which nitrogen is burned out of rotting biomass and sent back into the air, thereby causing a loss of nitrogen for future crops

ANNEX 3: LIST OF GREEN MANURE/COVER CROP SPECIES MENTIONED IN *RESTORING THE SOIL*

(Note: Many of these systems may presently be in use in additional places of which I am unaware.)

Common name	Scientific name	Gm/cc Systems	Countries where Systems are used
Alder	*Alnus* spp.	S117	India, China, Vietnam
Azolla	*Azolla pinnata*	S70	India, Southeast Asia
Bambara groundnut	*Vigna subterranea*	S3 S39 S69	Mali, other Sahelian nations Mali
Black bean (see lablab bean)			
Broad bean (see fava bean)			
Buckwheat	*Fagopyrum esculentum*	S113	Rwanda, Brazil
Calopo	*Calopogonium mucunoides*	S2	(Never)
Centrosema	*Centrosema pubescens*	none	
Chickpea	*Cicer arietinum*	S112	Ethiopia
Choreque	*Lathyrus nigrivalvis*	none	
Common pea	*Pisum sativa*	S110 S115	Peru Brazil, Uganda

Common name	Scientific name	Gm/cc Systems	Countries where Systems are used
Cowpea	*Vigna unguiculata*	S3 S23 S24 S25 S26 S27 S28 S29 S30 S32 S36 S64 S69 S90	Mali Honduras, Guatemala, El Salvador, Nicaragua, Panama, many African nations Zambia Cambodia Vietnam, Indonesia Laos Mozambique, Malawi Madagascar Mali, Burkina Faso, Mozambique Mozambique Thailand Honduras Mali Thailand, Uganda
Crotalarias (except Sunn hemp)	*Crotalaria spp.*	S63 S69	Brazil Indonesia
Desmodium	*Desmodium ovalifolium and D. trifolium*	S88	Honduras, Belize
Faidherbia	*Faidherbia albida*	S5	All across the Sahel
Fava bean	*Vicia faba*	S102 S109 S111	Mexico, Guatemala, Peru, Ethiopia, Vietnam Peru Mexico
Fonio	*Digitaria exilis*	S69	Mali
Forage peanut (see perennial peanut)			
Garbanzo bean (see chickpea)			
Gliricidia	*Gliricidia sepium*	S3 S4 S48	Mali Honduras, El Salvador Zambia
Grasspea	*Lathyrus sativa*	S114	Ethiopia
Green bean (see mung bean)			
Groundnut (see peanut)			

Common name	Scientific name	Gm/cc Systems	Countries where Systems are used
Horsepea (see lablab bean)			
Hyacinth bean (see Lablab bean)			
Indigofera	*Indigofera spp.*	S93	Vietnam
Inga	*Inga spp.*	S82	Guatemala, Honduras, Costa Rica, etc.
Jack bean	*Canavalia ensiformis*	S9 S43 S57 S58 S71 S72 S75 S76 S84 S86 S93 S95	Mozambique and Madagascar Cambodia Mexico, Brazil Honduras Mexico Mexico Paraguay, Honduras Nicaragua Costa Rica, Nicaragua Paraguay, Honduras Vietnam Vietnam, Mexico, Honduras, Mozambique
Lablab bean	*Dolichos lablab*	S9 S15 S16 S17 S18 S24 S36 S66 S73	Mozambique, Madagascar Honduras, Peru, Ethiopia, Kenya, Rwanda, Tanzania, Malawi, Mozambique Thailand Honduras Bangladesh Thailand Thailand India Philippines, Malawi
Leucaena	*Leucaena leucocephala, L. diversifolia*	S60	Benin, Ghana
Lima bean	*Phaseolus lunatus*	S61 S74	Mexico, Guatemala, Honduras, Nicaragua Malawi, Ethiopia
Lojy Be (see cowpea)			
Mother of cacao (see gliricidia)			

Common name	Scientific name	Gm/cc Systems	Countries where Systems are used
Mucuna	*Mucuna spp.*	S1 S11 S13 S14 S40 S41 S52 S53 S54 S55 S56 S78 S87 S89 S92	(Never) Benin Vietnam, Honduras Panama Guatemala, Ncaragua, Paraguay, Brazil, Malawi Honduras Mexico, Guatemala, Honduras, Belize Ghana, Benin, Kenya, Mexico Mexico Honduras Mexico, Costa Rica Brazil Paraguay Rwanda Vietnam
Mung bean	*Vigna radiata*	S32 S33 S34 S76	Mozambique Vietnam, Indonesia Vietnam Nicaragua
Peanut	*Arachis hypogaea*	S21 S32 S38 S69	Malawi, Zambia Mozambique Paraguay, most of Africa Mali
Perennial peanut	*Arachis pintoi*	S83 S85 S91	Costa Rica, Nicaragua Honduras, Costa Rica Paraguay, Brazil
Pigeon pea	*Cajanus cajan*	S6 S19 S20 S21 S22 S24 S96	Sudan Panama, Brazil, Zambia, Malawi Zambia, Malawi Malawi, Zambia Laos, Thailand Zambia Brazil
Piliostigma	*Piliostigma spp.*	S3 S10 S49 S50 S51 S65	Mali Ghana Burkina Faso Burkina Faso Mali Burkina Faso, Mali
Quickstick (see gliricidia)			

Common name	Scientific name	Gm/cc Systems	Countries where Systems are used
Rice bean	*Vigna umbellata*	S24 S35 S36 S37 S77	Zambia Mexico, Guatemala, El Salvador, Uganda, Kenya, Vietnam Thailand Vietnam Vietnam
Runner bean	*Phaseolus coccineus*	S97 S98 S99 S100 S101	Guatemala, Mexico, Honduras, Colombia, Ecuador, Peru, Bolivia, Chile Guatemala Kenya Highland Latin America Honduras
Scarlet runner bean (see runner bean)			
Sesbania	*Sesbania rostrata*	S68	Much of South and Southeast Asia
Soybean	*Glycine max*	S21 S59 S80	Malawi, Zambia China, Southeast Asia Vietnam
Stylo	*Stylosanthes spp.*	S44	Philippines
Sunn hemp	*Crotalaria juncea*	S62	Tanzania
Sweet clover	*Melilotus albus*	S103	Mexico
Tailcup lupine	*Lupinus caudatus*	S104	Ethiopia
Tarwi	*Lupinus mutabilis*	S106 S107 S108 S116	Peru, Ecuador, Bolivia Peru, Ecuador Peru Peru
Tephrosia	*Tephrosia vogelii* and *T. candida*	S3 S7 S8 S46 S47 S81 S98	Mali Honduras Cameroon, Thailand, Vietnam Honduras Haiti, Kenya, Vietnam, Indonesia, Philippines Guatemala Guatemala
Tithonia	*Tithonia diversifolia*	S12 S42 S45	Philippines Honduras Honduras

Common name	Scientific name	Gm/cc Systems	Countries where Systems are used
Velvet bean (see mucuna)			
Vetch	*Vicia spp.*	S105	Mexico
Wild sunflower (see tithonia)			
Yam bean	*Pachyrhizus erosus*	S79	Vietnam

ANNEX 4: SEED SOURCES FOR GREEN MANURE/COVER CROPS

In many cases, and more and more cases as time passes, seeds can be found in-country. Look for seeds in village markets, or those possessed by or accessible to NGOs with agriculture-related projects and research institutions (e.g., agricultural colleges and universities). If seeds of a particular crop or variety cannot be found in the country or region where you are working, it is possible to order seeds from other countries.

When transferring seeds across borders, you will likely need the following documents: 1) a seed import permit from the country that the seeds will be shipped to; 2) a phytosanitary certificate from the country of origin (often the shipper/seed supplier can supply this) stating that the seeds are free of pests and diseases. Check with the Ministry of Agriculture in the country you work in to find out details regarding documents needed and how to obtain them.

Typically, one would begin by obtaining the import permit, as this document could contain specific seed treatments required for entry of particular crops. One would then send the import permit to the seed supplier in the country of origin. Based on requirements outlined in the import permit, the seed supplier can carry out any specific seed treatments, list those in the phytosanitary certificate, and include both documents (originals may be required) in the box of seeds to be shipped. Request that shipments be sent via a reputable carrier such as DHL. Be aware that extra costs can be incurred if it becomes necessary to hire a separate courier to clear seeds through customs once they arrive in your country. Below are a number of seed suppliers that carry seeds of gm/ccs. Many of them can also provide you with certification that the seed they are sending is phytosanitary (ie free of insects and diseases).

Territorial Seed Company
PO Box 158
Cottage Grove, OR 97424
http://www.territorialseed.com/
Phone Orders: 800-626-0866
Fax Orders: 888-657-3131
Customer Service/Gardening Questions: 541-942-9547
Customer Service Email: info@territorialseed.com
Source of clover and hairy vetch

B & T World Seeds
Paguignan
34210 Aigues-Vives
France
www.b-and-t-world-seeds.com
Tel: 00 33 (0)4 68 91 29 63
Numerous species and varieties in stock; allow up to 8 weeks for seeds not in stock.

Baker Creek Heirloom Seeds
2278 Baker Creek Road
Mansfield, MO 65704 USA
www.rareseeds.com
Tel: (417) 924-8917
Fax: (417) 924-8887
Email: seeds@rareseeds.com
Source for a number of cowpea varieties

Banana Tree
715 Northampton St.
Easton, PA 18042 USA
https://banana-tree.com/
Tel: (610) 253-9589
Fax: (610) 253-4864
Email: faban@banana-tree.com
Source for jack bean

CIAT
International Center for Tropical Agriculture
Km 17, Recta Cali-Palmira
Apartado Aéreo 6713
Cali, Colombia
Tel: +57 2 4450000
Fax: +57 2 4450073
E-mail: ciat@cgiar.org
www.ciat.cgiar.org
URL for regional office contact info:
https://ciat.cgiar.org/where-we-work/

A CGIAR (Consultative Group on International Agricultural Research) research institution with an extensive collection bean and forage lines. URL for information on how to request seeds: https://ciat.cgiar.org/what-we-do/crop-conservation-and-use/

ECHO Seed Bank
17391 Durrance Road
North Fort Myers, FL 33917
Email: echo@echonet.org
www.ECHOcommunity.org
Trial packets of a number of gm/cc crops are available to ECHO's network of international development workers. See the website for information on how to join ECHO's network and request seeds.

Inland & Foreign Trading Co., LTD
Block 1090, #04-04/05
Lower Delta Road
Tiong Bahru Industrial Estate
Singapore 169201
Tel: +65 6272 2711
See legumes under "Legume Cover Crop" and "Pasture Seed" categories. Source of bulk quantities (50 kg).

Shivalik Seeds Corporation
47, Panditwari, P.O. Prem Nagar
Dehradun, 248007, Uttaranchal
India
TeleFax: +91 135 2773348
Email: hilander@vsnl.com
Source for a number of annual gm/ccs

Setropa B.V.
Troelstralaan 4
Postbox 203
1400 AE Bussum
Holland
www.setropa.nl
Tel: +31 (0)35 5258754
Fax: +31 (0)35 5265424
Email: setropa@setropa.nl
Source of a number of gm/ccs

Wolf and Wolf Seeds
Rua Paulo Padovan, 81
Ribeirão Preto-SP
Brasil
Tel: +55 16 2111 0505
Email: online form
www.wolfseeds.com
Source of gm/ccs in bulk amounts

Heritage Seeds Pty. Ltd
26 Prosperity Way
Dandenong South VIC 3175
Australia
www.heritageseeds.com.au
Tel: 03 9701 4000
Fax: 03 9701 4040
Email: heritage@heritageseeds.com.au
Source for several tropical and temperate legumes

If you are aware of other sources of seeds, please communicate that to the author at rbunchw@gmail.com and to the moderator of the Green Manure/Cover Crops discussion group on https://conversations.echocommunity.org/c/agricultural-techniques/GMCC

ANNEX 5: LATE-BREAKING NEWS

The following three gm/cc systems have come to our attention since the book was sent to the publisher, but we consider them important enough to be included in an Appendix:

The book provides no ideas as to gm/cc species that can be intercropped with maize when other low-lying plants, such as common beans, mung beans, rice beans, squash or sweet potatoes are intercropped with the maize. However, recent experience indicates that if pigeon peas are added to such systems where legumes or other low-lying crops are already intercropped with maize, these low-growing crops will grow perfectly well because of the open canopy of the pigeon peas. In a sense, this is merely an expansion of the doubled-up legume system (S21) to include almost any other low-stature legume, instead of limiting it to just soybeans or peanuts.

S118. Maize, sorghum or millet/pigeon pea/almost any other low-growing legume, such as mung beans or cowpeas, or even non-legumes, such as sweet potatoes or squash. Tanzania. Experimental. A handful of farmers. This system consists of intercropping maize, sorghum or millet with pigeon peas and almost any other low-growing crop that might also be intercropped with the taller traditional grains. This means that where maize is already intercropped traditionally with another crop at any elevation up to about 2,000 masl, pigeon peas can be added to the traditional system, providing more nitrogen, more biomass, more biodiversity and another high-protein food-crop to the traditional system at almost no extra cost. Thus, it could have major importance throughout huge areas of Central and South America and Africa.

The second system to be added here is one that could be used to establish certain perennial legumes in droughty, lowland areas where the presence of free-grazing animals has not allowed these legumes to be grown previously.

S119. The establishment of Gliricidia, Piliostigma, and other perennial gm/ccs. Mali. Experimental. A handful of farmers. A number of very good gm/ccs that we previously could only grow in droughty lowland areas with much difficulty because of the presence of free-grazing cattle, can be established quite easily, using this system in unimodal areas. These perennials are ones that can withstand free-grazing cattle any time except their first 16 to 18 months. Gliricidia, for instance, will grow out of the reach of cattle in 18 months even in areas of less than 400 mm/year of rain, and Piliostigma becomes unpalatable to cattle before 16 months from planting. These perennials can therefore be planted at the beginning of the rainy season, along with farmers' other crops. Then, just before the crops are harvested and the domestic animals are let loose, temporary fences of bushes and thorns can be built. The rains would return well before the fences fell apart, ensuring that the perennials were well-established before their second dry season begins. These fences, that are commonly made in such areas for growing small plots of irrigated vegetables, do take some work, but would definitely be worth it if farmers can establish patches of land that for the next 60 years or so could be farmed during the rainy season as the soil gradually became more and more fertile.

The third system is a traditional alley-cropping system that has been used for generations among Kenyan smallholder farmers just south of Mount Kenya.

S120. Maize, sorghum, or pearl millet/short-stature gm/ccs or lablab beans with alleys of pigeon peas. Kenya. Proven. Hundreds of thousands of farmers. This system consists of planting rows of pigeon peas every 4 to 6 m across a field of maize, sorghum or pearl millet. Then, additional low-growing legumes such as peanuts or cowpeas are intercropped with the cereals. Sometimes, lablab beans are intercropped with the maize in the alleys. This system has maintained very good levels of soil organic matter and resistance to drought.

I would suspect that in new areas, this system would be easier to introduce, and would produce nearly the same impact, if the alleys were 8 to 10 m wide, rather than just 4 to 6 m.

NOTES

Acknowledgements

[1]Primavesi, Ana, *Manejo Ecologico del Suelo, La Agricultura en Regiones Tropicales*, Fifth Edition (Buenos Aires: El Ateneo, 1984).

[2]Monegat, C., *Plantas de Cobertura del Suelo, Caracteristicas y Manejo en Pequenas Propriedades* (Chapeco, Brazil: Claudino Monegat, 1991).

Chapter 2: A Short History of Green Manure/Cover Crops

[3]*Ibid.* p 195.

[4]See, for instance, Colombo, C., *Use of Leguminous Plants in Tropical Countries as Green Manure, as Cover and as Shade* (Rome, Italy: Printing Office of the Chamber of the Deputies, 1936).

[5]Eilitta, M., *et al.*, eds, *Green Manure/Cover Crop Systems of Smallholder Farmers, Experiences from Tropical and Subtropical Regions* (Dordrecht, The Netherlands: Kluwer Academic Publishers, 2004), pp. 8-9.

Chapter 3: Green Manure/Cover Crop Systems

[6]Bationo, A., *et al.*, *Lessons Learned from Long-term Soil Fertility Management Experiments in Africa* (Springer Science+Business Media: Dordrecht, The Netherlands, 2012), p. 22.

[7]Binder, U., *Manual de Leguminosas de Nicaragua, Volume I* (Estelí, Nicaragua: PASOLAC and the Escuela de Agricultura y Ganadería de Estelí, 1997).

[8]Uphoff, N., *et al.*, "Soil Fertility as a Contingent Rather than Inherent Characteristic, Considering the Contributions of Crop-Symbiotic Microbiota," chapter 6 in R. Lal and B. A. Stewart, eds., *Principles of Sustainable Soil Management in Agroecosystems* (Boca Raton, Florida: CRC Press, 2013).

[9]This is now widely agreed upon now. See, for instance, Florentin, M., *et al.*, *Green Manure/Cover Crops and Crop Rotation in Conservation Agriculture on Small Farms* (Rome, Food and Agriculture Organization of the United Nations, 2011).

[10]National Academy of Sciences (NAS), *Tropical Legumes, Resources for the Future* (Washington, D. C.: NAS, 1979) and Tisdale, Samuel L., *et al.*, *Soil Fertility and Fertilizers*, Fifth edition (New York: MacMillan Publishing Co., 1993).

[11]Personal communication from Dr. David Johnson.

[12]These calculations were made by the author on the basis of biomass quantities reported in Florentin, *op. cit.*, and long-term soil sequestration rates reported by Dr. David Johnson.

[13]Scientific reports on this issue come to varying conclusions. Some, often using gm/ccs I would never recommend, claim gm/ccs dramatically increase labor needs and are therefore too expensive for smallholder farmers. Others, often using systems either recommended in this book or similar to them, report very low or negative costs. See, for instance, Bishop-Sambrook, C., *et al.*, "Conservation Agriculture as a Labour Saving Practice for Vulnerable Households," International Fund for Agriculture Development (IFAD) and Food and Agriculture Organization (FAO), 2004.

[14]Dyck, E., *et al.*, *Legumes in Every Farm, Strategies for Adoption of Soil-Improving Legumes by Smallholders in Kenya*, 1996 and Mureithi, J., *et al.*, eds., *Enhancing Agricultural Productivity in East Africa, Development and Up-Scaling of Green Manure Legume Technologies in Kenya* (Nairobi, Kenya: Kenya Agricultural Research Institute and Legume Research Network Project, 2006).

[15]Mureithi, J., *et al.*, "The Role of Green Manure Legumes in Smallholder Farming Systems in Kenya, The Legume Research Network Project," *Tropical and Subtropical Agroecosystems*, Vol. 1, 2003, p. 66.

[16]Kimenju, J. , *et al.*, "Use of green manure plants in cropping systems to suppress root-knot nematodes," *African Crop Science Conference Proceedings,* Vol. 8, pp. 2083-2085 and Mureithi, *Enhancing Agricultural Productivity in East Africa*, pp. 323,326.

[17]Eilitta, *op. cit.*, p. 23.

[18]Mureithi, *Enhancing Agricultural Productivity in East Africa*, p. 150.

[19]Buckles, D., *et al.*, *Cultivos de Cobertura en Agricultura de Laderas, Innovación Campesina con Mucuna* (Ottawa: Centro Internacional para el Mejoramiento de Maíz y Trigo and Centro Internacional de Investigaciones para el Desarrollo, 1998).

[20]Bunch, R., "El Trabajo de EPAGRI en el Estado de Santa Catarina, Brasil, Nuevas Posibilidades Importantes para Agricultores de Escasos Recursos." No date, unpublished.

[21]Flores, M., and N. Estrada, "Estudio de Caso, La Utilización de Frijol Abono (*Mucuna* spp.) Como Alternativa Viable para el Sostenimiento Productivo de Sistemas Agrícolas del Litoral Atlántico." Paper presented to the Center for Development Studies at the Free University of Amsterdam, 1992.

[22]*Ibid.*

[23]Personal communication from personnel of the Zamorano PanAmerican Agricultural School.

[24]Sekamattea, B.M., *et al.*, "Effects of maize-legume intercrops on termite damage to maize, activity of predatory ants and maize yields in Uganda," *Crop Protection*, Vol. 22, 2003, pp. 87–93.

[25]Lathwell, *op. cit.*

[26]Mureithi, "The Role of Green Manure Legumes in Smallholder Farming Systems in Kenya," p. 63.

[27]Triomphe, B., *Seasonal Nitrogen Dynamics and Long-term Changes in Soil properties under the Mucuna/Maize Cropping System on the Hillsides of Northern Honduras*, Dissertation for PhD, Cornell University, January 1996, pp. 110-122. This dissertation reports an average yield increase of from about 1.8 t/ha to 3.3 t/ha over 10 years, with most of that increase in the first few years. The base yield is probably after fallowing, because the average smallholder harvest of maize in Honduras was less than 1.3 t/ha.

[28]Eilitta, *op. cit.*, p. 123.

[29]I say this after having done dozens of economic analyses in the field, but other people increasingly agree. For instance, Benson said this of maize in Malawi back in 1997, when fertilizer prices were about a third what they are now. Also see Eilitta, *op. cit.*, p. 155. FAO calculations for cassava around the world show fertilizer is unprofitable at today's prices. I know from personal communications that much of the push in African countries for

governments to subsidize synthetic fertilizer is occurring because the people involved know that synthetic fertilizer does not pay at present farm-gate prices when used by many smallholder farmers.

[30]Florentin, *op. cit.*, p. 13.

Chapter 4: Improving Soil Fertility and Soil Moisture

[31]Primavesi, *op. cit.*

[32]For a much more complete explanation of this section on plant nutrition, see the paper Bunch, R., "Nutrient Quantity vs Nutrient Access? A New Understanding of How to Maintain Soil Fertility in the Tropics," which is available at http://edn.link/mjknct and on a number of other websites.

[33]*Ibid.*

[34]Eilitta, *op. cit.*, pp. 24-25.

[35]Triomphe, *op. cit.*, pp. 110-122.

[36]Committee on Tropical Soils, Agricultural Board, National Research Council, *Soils of the Humid Tropics* (Washington, D. C., National Academy of Sciences, 1972). Quotes are taken from pp. 65-66 and p. 71. This particular example was published before the gm/cc movement had come into existence. Other such sources continued publishing such information well into the 21st century. Furthermore, well before 1972, it was known that there were dozens of examples in dozens of countries in Latin American and Asia in which literally hundreds of thousand of farmers had been increasing or maintaining their yields of field crops for decades, and in some cases centuries, by using such gm/cc species as mung beans, rice beans, runner beans and tarwi.

[37]Malcolm Cresser, *et al.*, Soil Chemistry and its Applications (Cambridge, U.K., Cambridge University Press, 1993), p. 46.

[38]Personal communication with Yvonne Otieno, ICRAF's Communications Officer, in 2012.

[39]Bationo, *op. cit.*, p. 22.

[40]*Ibid.*, chapter 10. See graph, p. 192.

[41]Aihou, K., *et al.*, "Alley Cropping in the moist savanna of West-Africa: I.

Restoration and maintenance of soil fertility in 'terre de barre' soils in Benin Republic" in *Agroforestry Systems*, vol. 42, 1999, pp. 213-27.

[42]Reganold, J. and J. Glove, "A Cure for Africa's Soil," *Scientific American*, May 2016, vol. 314, no. 5.

[43]Nyang'oro, J., "Africa's Environmental Problems," in A. Gordon and D. Gordon, eds., *Understanding Contemporary Africa*, 3rd edition (Boulder, Colorado, USA: Lynne Rienner Publishers, 2001), pp. 230-231.

[44]Wagner, R., *Environment and Man*, quoted in Nyang'oro, *Ibid.*, p. 231.

[45]Bot, A. and J. Benitez, *The Importance of Soil Organic Matter, Key to Drought-resistance and Sustained Food Production* (Rome, Italy: Food and Agriculture Organization of the United Nations, 2005), chapter 5.

[46]Personal communication from Michael Hand.

[47]See, for instance, Kassam, A., *et al.*, "Overview of the Worldwide Spread of Conservation Agriculture," Facts Reports 8, Institut Veolia, *Field Actions Science Reports*, 2015.

[48]The Brazilians, with their far greater use of Conservation Agriculture than in Africa, totally agree with the idea that gm/ccs are vital to widespread adoption of CA. See Florentin, *op. cit.*, p. 73.

[49]Giller, K., *et al.*, "Conservation Agriculture and Smallholder Farming in Africa, The Heretics' View," *Field Crops Research*, Vol. 114, No. 1, pp. 23-34.

[50]A very interesting, though somewhat outdated, introduction to the biology of the soil is David W. Wolfe, *Tales from the Underground, A Natural History of Subterranean Life* (Cambridge, Mass.: Perseus Publishing, 2001).

[51]Brown, G. video of speech called "Keys to Building a Healthy Soil," at https://www.youtube.com/watch?v=9yPjoh9YJMk.

[52]Stoorvogel J. and Smaling, E., *Assessment of Soil Nutrient Depletion in sub-Saharan Africa 1983-2000*. Report no. 28, vol. 1-4 (Winand Staring Center: Wageningen, The Netherlands, 1990).

[53]Many scientists now agree with the idea that N is more limiting than P in African soils. For Kenya, see Mureithi, *op. cit.*, p. 125, and for Malawi, see Eilitta, *op. cit.*, p. 158. For Africa in general, see Buresh, R., *et al.*, eds., *Replenishing Soil Fertility in Africa, SSSA Special Publication No.*

51 (Madison, Wisconsin: Soil Science Society of America and World Agroforestry Center, 1997), p. 11.

[54]Stoorvogel and Smaling, *op. cit.*

[55]Sattari, S., *et al.*, "Residual Soil Phosphorus as the Missing Piece in the Global Phosphorus Crisis Puzzle," in *Proceedings of the National Academy of Sciences of the United States of America (PNAS)*, vol. 109, number 16, April 17, 2012.

[56]Triomphe, *op. cit.*

Chapter 5: Building Resilience to Droughts

[57]These calculations are based on information given to me in an informal paper and personal interviews with Dr. David Johnson.

[58]Rusinamhodzi, L., *et al.*, "Maize–grain legume intercropping is an attractive option for ecological intensification that reduces climatic risk for smallholder farmers in central Mozambique," *Field Crops Research*, vol. 136, 2012, pp. 12–22.

[59]Brown, *op. cit.*

[60]Personal communication from Bernard Triomphe. Bernard had the soil from some maize/mucuna sites analyzed at a soil lab in the United States. When the results showed 12%, he repeated the process, and the second analysis gave the same result. This level of SOM may sound impossible, but during a visit to Wageningen in the Netherlands, I visited the International Soil Reference and Information Centre (ISRIC). Personnel of ISRIC showed me a soil profile, from a tropical garden, in which the organic matter content was 12%.

Chapter 6. Green Manure/Cover Crop Species

[61]Taken from a video of a speech called "Increasing Soil Carbon Using Adaptive Multi-Paddock (AMP) Grazing," by Dr. Steven Apfelbaum, presented at a Perennial Farm Gathering, describing the results of the Minnesota Flood Plain Restoration Project, which is managed by a National Science team across the 50-km-wide Red River floodplain.

[62]Reganold and Glove, *op. cit.*

[63]Binder, *op. cit.*, p. 37.

[64]Mureithi, *Enhancing Agricultural Productivity in East Africa*, p. 127.

[65]Monegat, *op. cit.*, pp. 140-141.

[66]Kowal, T., "Manual sobre el Manejo y Aprovechamiento del Frijol Gandul," 1994.

[67]Mureithi, personal communication about trials carried out by the Kenyan government.

[68]Lampariello, L., *et al.*, "The Magic Velvet Bean of Mucuna Pruriens," *The Journal of Complementary and Traditional Medicine*, Volume 2, No. 4, October-December, 2012, pp. 331-339.

[69]Versteeg M., *et al.*, "Collaboration to increase the use of Mucuna in production systems in Benin," in D. Buckles, *et al.*, *Cover Crops in West Africa: Contributing to Sustainable Agriculture* (International Development Research Center (IDRC), International Institute for Tropical Agriculture (IITA) and Sasakawa-Global 2000: Canada, 1998), pp. 33-43.

[70]Personal communication from Tony Rinaudo.

[71]*Ibid.*

[72]See, for instance, Cherry, S., "Tephrosia fallow management in Cameroon," in *ECHO Development Notes, Issue #65*, 1999.

[73]Personal communication from Richard Dick by email.

[74]IFAD, *et al.*, *Shifting Cultivation: Towards Sustainability and Resource Conservation in Asia* (Cavite, Philippines: International Institute of Rural Reconstruction, 2001), pp. 238-243.

[75]Cairns, M., ed., *Voices from the Forest, Integrating Indigenous Knowledge into Sustainable Upland Farming* (Washington, D.C.: Resources for the Future, 2007).

[76]Guto, S., *et al.*, "Minimum Tillage and Vegetative Barrier Effects on Crop Yields in Relation to Soil Water Content in the Central Kenya Highlands, *Field Crops Research*, Vol. 132, no. 1, 2004, pp. 129-138.

Chapter 7. Achieving Adoption of Gm/cc Systems

[77]Bunch, R. and Kadar, A., "La Mucuna en los Sistemas de Agricultura de Bajos Insumos Externos en Mesoamerica," in *Revista LEISA*, vol. 20, no. 1, 2004.

[78]Bunch, R., "Africa's Soil Fertility Crisis and the Coming Famine," Chapter 6 in Worldwatch Institute, *State of the World 2011, Innovations that Nourish the Planet* (Washington, D.C.: Worldwatch Institute, 2011), pp. 59-68. The prediction about the drought could hardly have been more accurate. Famines in Africa, according to the accepted UN definition, officially started occurring (mostly in Nigeria, Somalia and Sudan), in early 2017. Where the prediction went wrong was in the actual mortality rate during 2017, because my knowledge of the potential size of food relief efforts was lacking.

[79]Ford, P., "How A 20-million-person Crisis Goes Unseen," *The Christian Science Monitor*, July 28, 2017.

[80]Bunch, R., "El Uso de Abonos Verdes por Agricultores Campesinos, Lo Que Hemos Aprendido Hasta la Fecha," *Reporte Técnico* No. 3, Segunda Edición, Tegucigalpa, Honduras: Centro Internacional de Investigación y Divulgación de Cultivos de Cobertura (CIDICCO), 1995.

[81]Bunch, R., *Two Ears of Corn: A Guide to People-Centered Agricultural Improvement* (Oklahoma City, Oklahoma, World Neighbors, 1982).

Chapter 8. Getting Started

[82]Cairns, *op. cit.*

Chapter 10. Choosing Lowland Farming Systems By Type and Food Preference

[83]See chapters 10 and 14 of Bunch, *Two Ears of Corn.*

[84]Guto, *op. cit.*

[85]See, for instance, Florentin, op.cit., pp. 10-11.

[86]Personal communication from Luís Alvarez Wélchez, technical director of the United Nations Food and Agriculture Organization's Lempira Sur Program.

[87]A very good report on a part of this study can be found in Cairns, M., *et al.*, "Shifting Forests in Northeast India; Management of *Alnus nepalensis* as an Improved Fallow in Nagaland," in Cairns, *op. cit.*, pp. 341-378.

[88]World Agroforestry Centre (ICRAF), Improved Fallows for Western Kenya, *An Extension Guideline* (Nairobi, Kenya: World Agroforestry Centre, 2003).

BOTANICAL INDEX

SUBJECT INDEX